Forage for pollinators in an agricultural landscape

Edited by Andrew Matheson & Norman L Carreck

Forage for pollinators in an agricultural landscape

Edited by
Andrew Matheson & Norman L Carreck

©2014 International Bee Research Association

ISBN 10: 0-86098-277-7

ISBN13: 978-0-86098-277-7

Published by and available from:
International Bee Research Association
6 Centre Court, Main Avenue,
Treforest, U.K. CF37 5YR

www.ibra.org.uk

Front Cover
Image: Norman Carreck.
Design: Joanne Hawker.

Contents

Abbreviations used

EC	European Commission
ecu	European currency unit (Euro)
EU	European Union
IBRA	International Bee Research Association
MAFF	Ministry of Agriculture, Fisheries and Food (now DEFRA, UK)
WW2	World War Two

Editorial Note: All illustrations have been contributed by the first author, except where otherwise noted.

Preface - Government and European Union land use policy

Liz McIntosh

Strategic Policy - Pollinators and Plant Health Safeguarding Plant and Animal Health, Department for Environment, Food and Rural Affairs, Sand Hutton, York, YO41 1LZ, UK

Bees and other insect pollinators have intrinsic cultural value, and play an essential role in the diversity and resilience of our plant and animal life, through the pollination of agricultural crops and wild plants. Although we know much about managed honey bees, evidence about the current status of our other pollinators is poor, and we know little about the value of the pollination services provided by insects. There is, however, a growing consensus that the many threats pollinators face may have caused declines in abundance, diversity, and geographical range of some species.

Defra's National Pollinator Strategy, to be launched in 2014, aims to cover all of the approximately 1,500 insect species that fulfil a pollination role in England. It aims to ensure that pollinators thrive and provide essential pollination services for agriculture and the wider environment, through addressing gaps in our evidence base, while developing new actions on the basis of our improved knowledge.

The Strategy identifies action in six core areas: 1. A 'Call to Action' message for bees and other pollinators, providing a simple, balanced message to inform the public and land managers on how to fulfil the essential requirements of bees and other pollinators; 2. Evidence-based pollinator-friendly management of farmland, including integrated pest management; 3. Evidence-based pollinator-friendly management of towns and cities including public land, corridors along road and rail networks and private gardens; 4. Evidence-based and precautionary responses to pest and disease risks; 5. Improving understanding on the status of pollinators and the service they provide; and 6. Sharing knowledge and engaging the public.

Bees and other pollinators have specific needs to ensure their survival: access to food supplies (nectar and pollen); and places to shelter and nest, including over-wintering sites. The Government aims to reflect these essential needs in a simple message and associated advice. Farmland makes up around 70% of England, so agricultural land management is one of the most important influences on our biodiversity and ecosystem services, including the status of our pollinators. Agri-environment schemes through the Common Agricultural Policy (CAP) have already shown benefits in terms of supporting diversity and abundance of pollinators. With the ongoing reform of the CAP, this is a time of transition for farming policy. The Government is currently developing new CAP plans for implementation in 2015 and is considering how new policies on Greening, farming and forestry can provide opportunities for pollinators.

Urban and suburban landscapes can also provide a range of different habitats with the potential to cater for the needs of pollinators. Actions on management of this land will build on existing policies and initiatives and include policies on habitat and species conservation, and initiatives by businesses and non-government organisations.

There is a high level of public awareness on the issue of pollinators. To maximise the benefit of this awareness and to turn it into effective actions to help pollinators, Government will play a coordinating role, encouraging collaborative action that continues to be adapted to up-to-date evidence and is supported by consistent messaging.

Introduction

Norman L Carreck

International Bee Research Association, 6, Centre Court, Main Avenue, Treforest, CF37 5YR, UK
and
School of Life Sciences, University of Sussex, Falmer, Brighton, East Sussex, BN1 9QG, UK.

Mainly due to problems such as "Colony Collapse Disorder" experienced by beekeepers in recent years, the conservation of bees and other pollinators is currently very high on the public agenda. It may therefore now seem inconceivable that just twenty years ago, bee conservation was sadly not fashionable, and it was difficult to attract the interest of the public and politicians in the issue, and more seriously, difficult to attract funding to study the problems. From its earliest years, the International Bee Research Association (IBRA) flew the flag for bee conservation. The original Bumble Bee Distribution Maps Scheme was set up through the auspices of IBRA, and resulted in the pioneering work of Paul Williams which for the first time documented declines in many of our bumble bee species[11,12]. These studies demonstrated that many of our rarer bumble bees had retreated to remote coastal locations, whilst a "Central Impoverished Region" where only six common species occurred, had opened up in England. This region comprised the main arable area of the country, where extensive land use changes had taken place since the Second World War. These changes were originally aimed at improving the efficiency of food production, but through measures such as the removal of hedgerows, fertilisation of grassland and widespread herbicide use, they had dramatic effects on the food availability for bees and other pollinators, and on the availability of nest sites for many bee species.

Ironically, the very success of these improvements in agricultural efficiency let to new problems, with food shortages replaced by food surpluses. Measures intended to reduce overproduction through the policy of "set-aside", whereby land was taken out of production, seemed to provide possibilities for reversing some of this impoverishment of our countryside for the benefit of wildlife, including bees. The UN Conference on Environment and Development, held in Rio de Janeiro, Brazil in 1992 also moved biodiversity up the global political agenda.

This book thus resulted from a conference entitled "Pastures new" organised by IBRA, and held at Churchill College Cambridge, UK on 4 December 1993. Seven speakers considered the forage available to honey bees, bumble bees and solitary bees in farmland, and how new opportunities for improving forage for them might occur in this changing farmland.

In his foreword to the original book[8] (fig. 1), the then IBRA Director, Andrew Matheson wrote: "IBRA has acted to stimulate discussion on these important issues. We want beekeepers and bee scientists to be aware of the need to work for bee conservation, and of the opportunities that have been created by current agricultural policies. We also want agriculture industry professionals to consider, and adopt, land management strategies that will promote the availability of habitats and food sources for bees".

This was followed up by a two day symposium organised by IBRA and held at the Linnean Society, London, UK in April 1995 which brought 21 scientists from nine countries together with 100 other delegates to discuss a range of topics related to bee conservation. The resulting monograph[9] (fig. 2) is sadly now out of print. Three years later IBRA hosted the First European Workshop on Habitat Management for Wild Bees and Wasps held at the University of Cardiff, UK. The accompanying book[6]

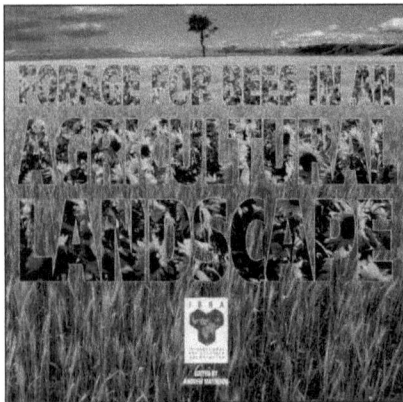

FIG. 1 . **Forage for bees in an agricultural landscape (1994).**

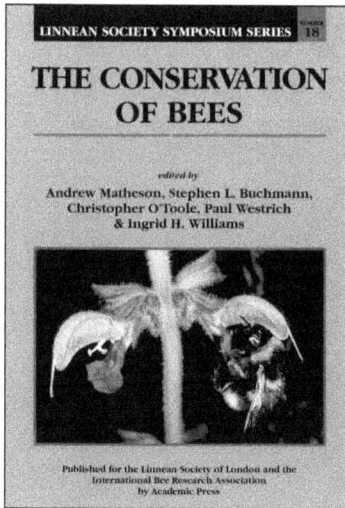

FIG. 2 . The conservation of bees (1996).

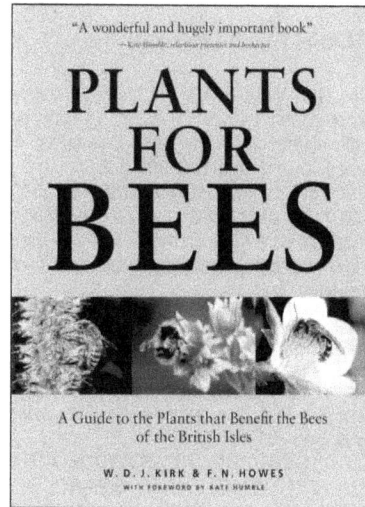

FIG. 4. Plants for bees (2012).

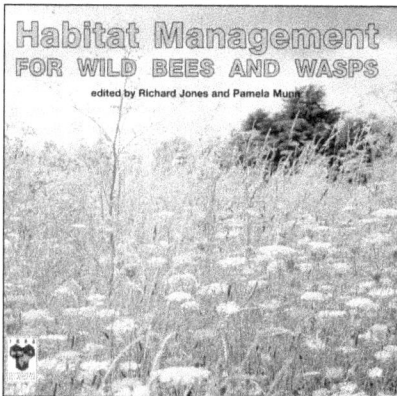

FIG. 3. Habitat management for wild bees and wasps (1998).

(fig. 3) is at the time of writing still available from the IBRA bookshop.

At the time of the original publication of this book, no trials of forage mixtures specifically intended for bees had been carried out in the UK, but as a direct result of the work described here by Wolf Engels, trials later took place of both annual[1,2] and perennial mixtures[3,4,5,10] of plant species, and "pollinator mixes" subsequently became options in the various stewardship schemes which replaced the old set-aside schemes.

An important contribution to the debate about suitable plant species for providing forage for bees has also been "Plants for bees"[7] (fig. 4) by William Kirk, published by IBRA in 2012, which built on the well-loved book by F N Howes written in the 1940s, and provides information for the gardener or farmer on suitable planting choices to be made.

As outlined by Liz McIntosh, the UK Government's new National Pollinator Strategy will link existing schemes for the management of agricultural land with broader policies to encourage all land holders to consider pollinators in their management strategies. It is therefore appropriate for IBRA to republish this important book, which has been unavailable for some time, in 2014 to coincide with this interest in the conserving our pollinators.

References

1. CARRECK, N L; WILLIAMS, I H (1997) Observations on two commercial flower mixtures as food sources for beneficial insects in the UK. *Journal of Agricultural Science, Cambridge* 128: 397-405.
http://dx.doi.org/10.1017/S0021859697004279

2. CARRECK, N L; WILLIAMS, I H (2002) Food for insect pollinators on farmland: Insect visits to the flowers of annual seed mixtures. *Journal of Insect Conservation* 6: 13-23.
http://dx.doi.org/10.1023/A:1015764925536

3. CARRECK, N L; WILLIAMS, I H; OAKLEY, J N (1999). Enhancing farmland for insect pollinators using flower mitures. *Aspects of Applied Biology* 54: *Field margins and buffer zones – Ecology, management and policy*. 101-108.

4. CARVELL, C; OSBORNE, J L; BOURKE, A F G; FREEMAN, S N; PYWELL, R F; HEARD, M S (2011) Bumble bee species' responses to a targeted conservation measure depend on landscape context and habitat quality. *Ecological Applications* 21(5): 1760-1771.

5. HEARD, M S; CARVELL, C; CARRECK, N L; ROTHERY, P; OSBORNE, J L; BOURKE, A F G (2007) Landscape context not patch size determines bumble bee density on flower mixtures sown for agri-environment schemes. *Biology Letters* 3: 638-641. http://dx.doi.org/10.1098/rsbl.2007.0425

6. JONES, H R; MUNN, P A (Eds) (1998) *Habitat management for wild bees and wasps*. International Bee Research Association; Cardiff, UK. 38 pp. ISBN 0 86098 235 1

7. KIRK, W D J; HOWES, F N (2012) *Plants for bees*. International Bee Research Association; Cardiff, UK. 311 pp. ISBN 978-0-86098-271-5

8. MATHESON, A (1994) *Forage for bees in an agricultural landscape*. International Bee Research Association; Cardiff, UK. 75 pp. ISBN 0 86098 217 3

9. MATHESON, A; BUCHMANN, S L; O'TOOLE, C; WESTRICH, P; WILLIAMS, I H (Eds) (1996) *The conservation of bees*. Linnean Society of London / International Bee Research Association / Academic Press; London, UK. 254 pp. ISBN 0-12-479740-7

10. PYWELL, R F; MEEK, W R; HULMES, L; HULMES, S; JAMES, K L; NOWAKOWSKI, M; CARVELL, C (2011) Management to enhance pollen and nectar resources for bumble bees and butterflies within intensively farmed landscapes. *Journal of Insect Conservation* 15: 853-864. http://dx.doi.org/10.1007/s10841-011-9383-x

11. WILLIAMS, P H (1982) The distribution and decline of British bumble bees (*Bombus* Latr.). *Journal of Apicultural Research* 21(4): 236-245.

12. WILLIAMS, P H; (1986) Environmental change and the distributions of British bumble bees (*Bombus* Latr.). *Bee World* 67(2): 50-61.

Land use changes and honey bee forage plants

Ingrid Williams[1]; Norman L Carreck[2]

[1]AFRC Institute of Arable Crops Research, Rothamstead Experimental Station, Harpenden, Herts. AL5 2JQ, UK

[2]International Bee Research Association, 6, Centre Court, Main Avenue, Treforest, CF37 5YR, UK
and
School of Life Sciences, University of Sussex, Falmer, Brighton, East Sussex, BN1 9QG, UK.

The United Kingdom has a diverse agriculture which supplies a variety of high quality food and other products. Agricultural land is integrated within an ecosystem of semi-natural vegetation which supports a varied wildlife. Here we examine why bees are an essential part of this agroecosystem, what types of bee are in the UK, and what they need to survive within this environment. We also examine changes in land use that have occurred since the Second World War and their implications for bees.

Why do we need bees?

Bees are needed to pollinate at least 38 different crops grown in the UK[43, 44]. The extent to which plants are dependent on bees for pollen transfer is determined by the structure of the flowers, their self-fertility and their arrangement on the plant or plants[9, 44]. Pollination by bees is essential to the 14 crops which are largely self-incompatible, for one of several reasons: their flowers are incapable of being fertilized by their own pollen (e.g. apple and pear); or they are monoecious, with male and female flowers on the same plant (e.g. gherkin, marrow and pumpkin); or are dioecious, with male and female flowers on different plants (e.g. kiwifruit); or are dichogamous, with male and female organs becoming effective at different times (e.g. fennel and coriander).

Bees are important pollinators of many crops with bisexual flowers which are partially self-incompatible (e.g. field bean, turnip rape, black and white mustard and plum), and for those with heterostylous flowers requiring pollen transfer between two flower forms (e.g. buckwheat), or those which are protandrous, with pollen release before stigma receptivity (e.g. sunflower). Bees are beneficial to plants with self-fertile but not totally auto-pollinating flowers, as their visits ensure pollen transfer within the flower (see table 1). About 32 plants grown as vegetable crops, for green manure or essential oil, depend on bees for the production of seed for their propagation (table 2).

There are varied benefits to be derived from pollination by bees, such as increased fruit or seed production (clover), improved fruit quality (strawberry), synchronized seed ripening (oilseed rape), improved oil content of seed (sunflower), and increased hybrid vigour with better germination and seedling establishment.

As well as being important for the pollination of crops, bees are undoubtedly important for the pollination of many of our native wild plants[7]. Much less is known about their pollination requirements but these plants in turn support other wildlife, for example the larval stages of the moths and butterflies that feed on their vegetation and the many small mammals and birds that feed on their seeds. Thus bees are a key group within both the agricultural and natural environments.

What bees are there in the UK?

In the UK, there are three types of indigenous bee: the solitary bees, the bumble bees and the honey bee.

Solitary bees

The role of the 250 species of solitary bee in pollinating crops and wild flowers in this country has been little studied[29]. They are generally far less numerous on crops than the bumble bees or honey bees, largely because their nesting habitats are discontinuous or patchily distributed. A few however, like *Osmia rufa*, can be common and locally abundant, and because they fly early in the season can be important pollinators of early flowering crops such as tree fruit (top fruit)[37]. Techniques for managing solitary bees for the pollination of some crops are being developed in continental Europe[46] and North America[34, 35].

Bumble bees

Bumble bees, of which we have 17 species of true bumble bees and six species of cuckoo bumble bees, have been more extensively studied. Although their distributions and abundance are declining in Britain[49], particularly in intensively cultivated areas, they can still be more abundant on some crops and wild flowers than honey bees. Methods for the continuous rearing of *Bombus terrestris* have now been developed and this species is being extensively used in glasshouses for the pollination of tomato crops[15].

8

TABLE 1. UK crops grown for their fruit or seed that need bees for pollination.		
Largely self-incompatible monoeocious, dioecious or dichogamous	Partially self-incompatible heterostylous or protandrous	Self-fertile but not auto self-pollinating
Bees essential	Bees important	Bees beneficial
apple	black mustard	aubergine
blackcurrant	broad bean	blackberry
caraway	buckwheat	dewberry
celery	field bean	gooseberry
coriander	plum	grapevine
fennel	raspberry	loganberry
gherkin	sunflower	morello cherry
kiwifruit	turnip rape	okra
marrow	white mustard	peach
melon		peppers
pear		redcurrant
pumpkin		strawberry
sweet cherry		swede rape
sweet fennel		tomato
		whitecurrant

Because of their long tongues and ability to work in poor weather conditions the importance of bumble bees as pollinators in the UK should not be underestimated.

Honey bees

The honey bee remains, at present, the only managed pollinator for field crops and is being increasingly relied upon as the distribution and abundance of wild bees diminishes. Field crops are generally grown in large monocultures, producing many millions of flowers per hectare, and locally available pollinators are often insufficient to pollinate all the flowers. The numbers of beekeepers and colonies have declined in recent decades and we now have an estimated 31 234 beekeepers and 170 223 colonies[3].

Most UK beekeepers are amateurs: only 1% of them own more than 40 colonies of bees. Most amateur beekeepers keep their bees in one place close to their homes, while commercial beekeepers frequently move their colonies from apiaries to crops to take full

TABLE 2. UK crops that need bees for propagative seed production.

Vegetable crops

asparagus	chives	radish
basil	endive	rocket
beetroot	marjoram	rosemary
cabbage	onion	salsify
carrot	pakchoi	spearmint
cauliflower	parsley	sugar beet
chervil	peppermint	turnip
chicory	petsai	

Green manure/forage crops

alsike clover	sainfoin	red clover
lucerne	vetch	white clover

Essential oil crops

borage	evening primrose	lavender

advantage of the nectar produced or to pollinate the crops[48].

What do bees need for survival?

Our indigenous bees have two essential requirements for survival: places to nest and flowers on which to forage. The nest sites and food plants of the solitary and bumble bees are discussed elsewhere[12, 13, 14, 28, 29]. The natural nest for the feral honey bee is a hollow deciduous tree but most honey bee colonies in the UK live in hives provided by beekeepers. The honey bee's main requirement from its environment is for a succession of flowering plants to supply nectar and pollen from spring through to autumn within foraging range of their hive. Estimates for the distances that honey bees will fly for food vary, but if we take it to be about 2 km the foraging range covers 1 256 ha of land. The number of colonies that can be kept profitably in a given area depends largely on the abundance of bee forage in that area. As colony density increases, competition for forage also increases and honey production per colony can decrease. There has been, in recent decades, a progressive urbanization of beekeeping as beekeepers, previously distributed throughout the countryside, have concentrated in urban or suburban areas. Here the floristic diversity and successional forage provided by gardens substitutes for that which previously characterized the countryside.

On which plants do bees forage?

To identify the main sources of food for honey bees in the 1930s to 1950s we examined some of the writings of eminent beekeepers and scientists of the time[16, 17, 18, 22, 23, 24]. Deans[8] made a nationwide melissopalynological survey in 1952, in which he identified the pollen grains in over 900 samples of honey: this remains the only objective study of the forage sources of honey bees in the UK.

At Rothamsted records have been analysed from a survey in which beekeepers reported the main nectar yielding plants in their area[48], and the findings were compared with earlier records. Some 130 different species or genera from agricultural and semi-natural environments were identified as important food sources for honey bees. Ornamental and garden plants are omitted from consideration here. Of these 130 plants, relatively few are major sources of nectar, that is, those which produce sufficient nectar to provide a honey harvest for beekeepers (table 3). The remainder are secondary sources of nectar, which can be important where abundant. We have grouped these 130 plants into the five different habitats in which they grow: crop plants; herbaceous 'weeds' of cultivated or uncultivated farmland; grassland or pasture plants; trees, shrubs and herbs of woodland and hedgerows; and the plants of inland and coastal moorlands and marshes.

We then examined the changes that have occurred to these habitats over the past five decades. There has been little overall change in the areas of land in arable production and under permanent pasture over the period 1945–1992, and only a slight decrease in total agricultural land and in rough grazing (fig. 1). But there have been major changes in the quality of this agricultural land as a source of forage for bees.

Crop plants

Cereals are the dominant arable crops, particularly in eastern England. Cereal production increased in the 1960s largely due to improved agronomic methods (fig. 2), and by 1992 cereals were grown on 56% of the 5.7 million hectares of land in arable cultivation in England and Wales. Modern weed-free cereal crops provide no food for bees except sometimes honeydew from aphids infesting the crop.

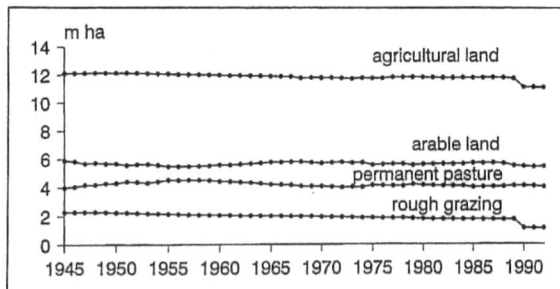

FIG. 1. Agricultural land use changes in the UK 1945–1992 (source: UK Ministry of Agriculture, Fisheries and Food (MAFF)).

10

Plant	Herrod-Hempsall [16]	Manley [22,23,24]	Howes [18]	Deans [8]	Hodges [17]	Williams et al. [48]	Plant type[b]
TABLE 3. Major nectar sources in the UK.							
Crops							
legumes							
clovers (white, alsike, red, crimson)	*	*	*	*	*	6	p
sainfoin	*	*	*		*		p
lucerne	*						p
field bean		*	*	*	*	7	p
fruits							
tree (apple, pear, plum, cherry)	*		*	*	*	3	p
soft (blackcurrant, raspberry, gooseberry, strawberry)	*				*		p
other							
Brassica sp.	*		*	*	*	1	a
buckwheat	*		*				a
Heathers							
ling heather		*	*		*	}5	p
bell heath			*		*		p
Trees/shrubs							
sycamore			*	*	*	8	p
lime			*	*	*	2	p
hawthorn			*		*	10	p
sweet chestnut						}9	p
horse chestnut							p
ivy						12	p
privet				*			p
blackberry			*	*	*	4	p
Herbs							
charlock		*	*		*		a
rosebay willow-herb		*	*		*	11	p
dandelion			*		*	14	p
balsam						13	p

[a]Numbers indicate the importance as a principal honey flow reported by Williams et al.[48], * = listed by author
[b]p = perennial; a = annual

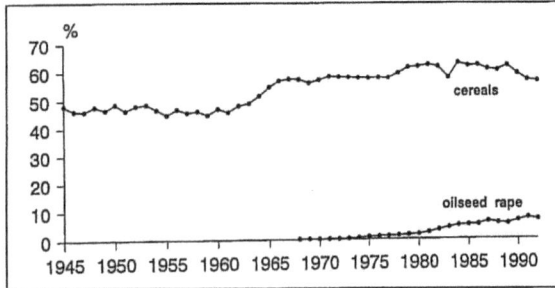

FIG. 2. Cereals and oilseed rape as a percentage of total arable area (source: MAFF).

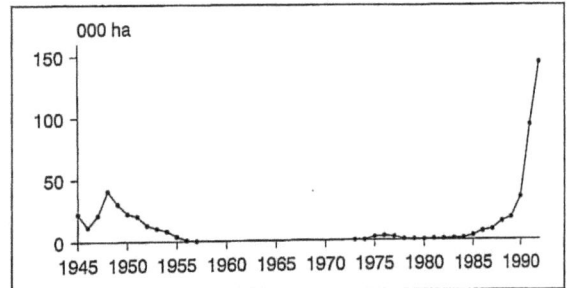

FIG. 3. Area of linseed grown, 1945–1992 (source: MAFF).

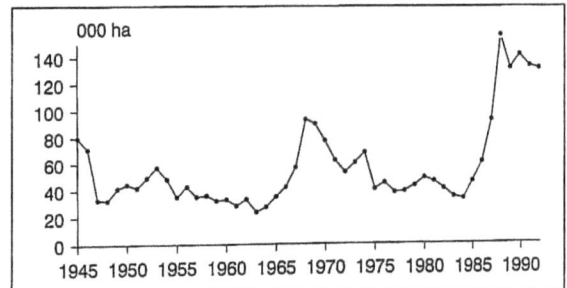

FIG. 4. Area of field beans grown, 1945–1992 (source: MAFF).

In the late 1960s, as cereal production increased, oilseed rape was introduced and now occupies almost 7% of arable land, making it the second largest crop in the country. The introduction of oilseed rape has been advantageous for honey bees as it provides an abundance of nectar and pollen, albeit for only a short period. This crop is extremely attractive to bees[38, 39], and in terms of potential honey yields it is more productive than any other UK crop plant[45]. Provided that pesticides toxic to bees are not applied to the crop while it is in flower, oilseed rape nectar can be the main contributor to the spring honey flow, replacing in many areas the summer flows previously obtained from semi-natural vegetation. We have estimated that the honey potential of rape is worth £40 million annually.

Linseed is another oilseed crop which has seen increased production recently (fig. 3). Linseed and flax are the same plant species, but linseed cultivars are grown for seed whereas flax cultivars are grown primarily for fibrous stems. In the UK flax production for rope manufacture was important during World War 2 (WW2) but had ceased by the mid-1950s. However, linseed production, started in the early 1970s, has increased rapidly to its present level of 142 100 ha because of recent EU support. Linseed is visited by honey bees and has been reported to provide a surplus honey flow for beekeepers in some years[41].

Field bean production has been variable since WW2, ranging between 23 300 and 139 200 ha, but has increased in recent years thanks to better agronomy, particularly disease control (fig. 4). Lucerne was a legume crop of great importance to honey bees up to the early 1970s but is no longer grown to any appreciable extent (fig. 5).

Between 1947 and 1992 most (70%) of our older orchards were taken out of production (fig. 6), a practice supported by government grants. An important spring food source for bees has thereby been reduced, but the figures for the area occupied by orchards give an exaggerated impression of the impact on bee forage because traditionally orchards were wood pasture with large trees, widely spaced, whereas today the trees are smaller, densely packed, more productive and have a higher flower density. The 1945–1946 production of apples was 323 000 tonnes whereas in 1992 it was 249 000 tonnes: only a 23% drop in production despite a corresponding 64% reduction in area.

The area of soft and bush fruit (berry fruit) is 16% less than the post-WW2 level mainly because of a reduction in the growing of raspberries and gooseberries (fig. 7). Production of broad beans and runner or

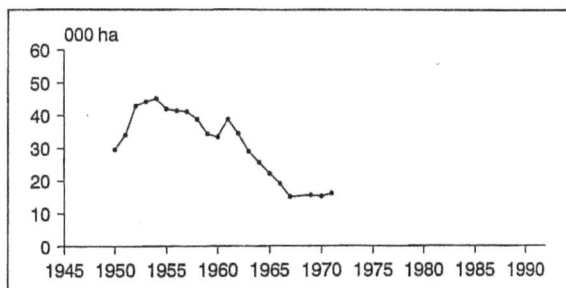

FIG. 5. Area of lucerne grown, 1945–1992 (source: MAFF).

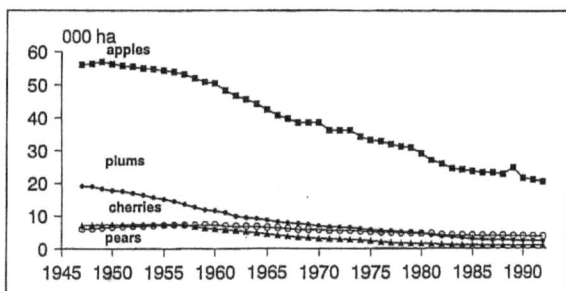

FIG. 7. Area planted in bush and soft fruit (berry fruit), 1945–1992 (source: MAFF).

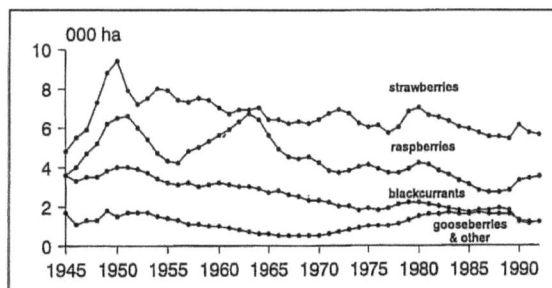

FIG. 6. Area planted in tree fruit, 1945–1992 (source: MAFF).

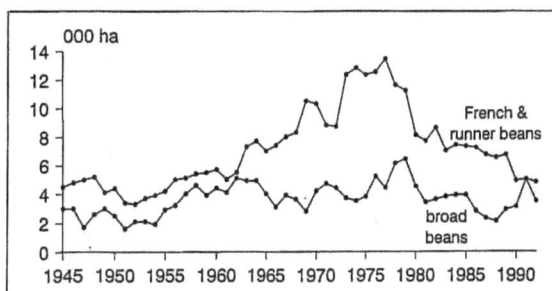

FIG. 8. Area of bean crops, 1945–1992 (source: MAFF).

French beans has either remained steady or increased slightly (fig. 8).

For bee forage from crop plants, the increase in area of oilseed rape has probably more than offset any area losses of the less extensively grown crops. But valuable as the crop sources of nectar and pollen may be, crops are grown in monocultures or discrete blocks, flower for a short period only and do not provide the succession of flowers that bees need to survive. Further, the increase in intensive arable farming at the expense of mixed farming has in itself reduced the diversity of forage available to a honey bee colony within its foraging range.

Herbaceous plants of cultivated and uncultivated farmland

Another practice that has decreased the successional availability of flowering plants for bees in agro-ecosystems has been the increased use of herbicides in cultivated areas, field margins, hedgerow bottoms and waysides.

The effect of different crop husbandry practices on the weed infestation of cereals is very evident in the most famous of the classical experiments at Rotham-sted, the Broadbalk experiment, established in 1843. In this experiment winter wheat has been grown continuously on the same land for 150 years. Plots have been variously treated with organic manures or inorganic fertilizers supplying the elements N, P, K, Na and Mg in various combinations. Some plots have received no herbicides. About 50 annual and 10 perennial weed species occur in the field. Some species such as corn buttercup occur in all plots, but others are associated with manurial treatments (e.g. legumes where minerals but not nitrogen are applied).

By contrast, herbicide and pesticide treated plots can have fewer than five species, represented by only one or two plants of each. Herbicides then, have greatly decreased plant diversity in our cereal fields. Weeds

FIG. 9. The Park Grass experiment at Rothamsted, showing the diverse bee flora that will grow on pasture receiving no nitrogen.

such as common vetch, corn buttercup, black medick, mayweed, charlock, cornflower, knapweed, poppy, scabious, thistles and speedwell no longer provide bees with a succession of flowers[33]. Weed control and seed cleaning are now so efficient that some former weeds such as corncockle, cornflower and field cow-wheat are now endangered plant species, while others such as corn buttercup and poppies are seldom seen. Others including corn marigold have declined through the use of lime on acid soils[19, 30, 31].

Grassland or pasture plants

Changes in the management of grassland have likewise had a major impact on bee forage. In 1945, permanent grassland was usually rich in clovers and other broadleaved species. Clover was the single most important and widespread source of nectar for bees in the UK and much of our honey was predominantly clover honey. Although much (63%) of the utilized agricultural area is still permanent grassland, most (> 90%) has now been 'improved' by reseeding, is treated with herbicides and fertilizers, or is cut for silage before herbaceous plants have flowered. Short term leys, which in 1945 were usually grass/clover mixtures, are now usually Italian ryegrass alone. Few of

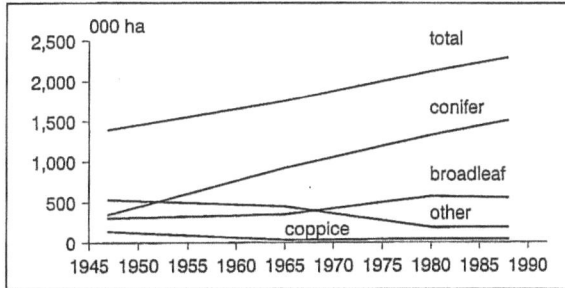

FIG. 10. Area in woodland, 1947–1988 (source: Forestry Commission).

the floristically diverse old river meadows or chalk-grasslands remain.

Another of the classical experiments at Rothamsted, the Park Grass experiment, established in 1856 (fig. 9), demonstrates in a unique way how continued manuring with different organic and inorganic fertilizers applied to plots maintained at different pH levels affects the botanical composition and the yield of a mixed population of grasses, clovers and other herbs[33, 36]. Unmanured plots, which approximate most closely to the state of the whole field in 1856, have the most diverse flora with 50–60 species, with many bee forage plants including meadow vetchling, birds-foot trefoil, red clover, white clover, yarrow, common knapweed, rosebay willow-herb, field scabious, chick-weed, dandelion and speedwell. Farmyard manure has maintained the proportion of legumes but reduced other plants, whereas application of inorganic nitrogen has increased grasses at the expense of broadleaved species especially legumes, a balance that could to some extent be redressed by liming.

In the past it was mainly the uncultivated areas within the agricultural landscape that provided the succession of flowers for bees. But areas such as field margins, hedgerows, copses, waysides, water courses, ponds and wet areas have gradually been lost.

Woodlands and hedgerows

British woodlands have undergone major changes in recent decades[27] (fig. 10). The total amount of woodland has almost doubled since WW2, when timber reserves went towards the war effort, and now occupies about 9% of our total land area. The type of woodland has changed considerably: most of the commercial planting over the past three decades has been of fast-growing conifers rather than of broadleaved trees (at the rate of c. 20 000 ha/year), and the amount of coppice and farm woodland or scrub has declined.

There is a common belief that broadleaved woodland is good for wildlife whereas coniferous plantations are bad. In terms of bee forage this is to a large extent true. All three layers of a deciduous woodland can provide nectar and pollen. Trees like sycamore, lime, chestnut, oak, elm and hawthorn are all visited by bees when in flower. The shrub layer can contain bramble, dog rose or ivy, and herbaceous plants can be plentiful in the ground layer, particularly in less densely shaded areas, in clearings or along rides. Coniferous woodland, by contrast, appears dark and inhospitable with little ground flora. Many conifers can of course yield honeydew, but few provide pollen or nectar. However, even upland spruce plantations can have valuable bee plants during the early stages of timber production, along the rides and in clear felled areas. Plants such as rosebay willow-herb, thistles, bilberry, heather, bramble, raspberry, holly, gorse, alder and birch can be encouraged by environmentally sensitive management.

The extent of hedgerows, which can be thought of as elongated strips of woodland, has also declined. In 1947, there were approximately 800 000 km of hedgerows but by 1990 almost half (47%) had been removed, largely aided until about 1970 by government grants. Losses have been greatest in the arable areas of East Anglia, some parts of which have lost all their hedges, and least in the pasture of the south-west.

Further losses still occur through neglect: without proper maintenance, which is expensive, hedges soon become derelict, damaged or reduced to a row of relic trees. Between 1984 and 1990 a net loss of 47 000 km of hedgerow (22% of the total) is estimated to have occurred. The Dutch elm disease of the 1970s virtually eliminated elm trees, an important landscape feature as well as a source of nectar and pollen. Loss of hedges means reductions in the abundance of

FIG. 11. Well-managed roadside verges can provide both nest sites and food for bees.

FIG. 12. Increased mechanization, larger fields and use of agrochemicals has reduced nest sites and food for bees in farmland.

hazel, alder, blackthorn, holly, hawthorn, bramble, sycamore, lime, and elms: all important sources of food for bees, as well as the ground flora associated with hedgerows.

Plants of moorlands, heaths, commons and marshes

Heather-dominated moorland has a special importance to beekeeping in this country, and beekeepers will move their colonies several hundred kilometres to glean the heather harvest. Heather moorland, although still extensive, is locally threatened by forestry and agriculture and more widely by encroachment of bracken, over- or under-grazing and inappropriate burning[21]. Losses of heather moorland in northern England, that is, Cumbria, North York Moors and the Peak District, are currently estimated to be 1% per annum. However, there is much research and conservation interest in the restoration of heathland ecosystems. Now management systems such as fertilization regimes, turfstripping and propagule collection and redistribution are being developed[26]. Heathland, that is heather dominated lowland areas with a mineral- as opposed to the peat-based soil of upland moorland, is probably the most threatened habitat. England now has only 10% of the heathland present 200 years ago. It faces losses from forestry, lack of

grazing, 'reclamation' for agriculture, or urban development[32].

In coastal areas salt marshes (total area 44 370 ha) can provide good bee forage, particularly from sea lavender, sea aster and thrift, and extensive areas of this type of habitat still exist, for example on the north Norfolk and north Kent coasts and in Chichester harbour.

Finally, another habitat of immense importance to bees is the one that criss-crosses all cultivated and uncultivated areas: the verges of roads (fig. 11) and railways. In the UK roads have a verge area of about 156 000 ha, which provides a variety of different habitats from permanent unfertilized grassland to secondary woodland and supports a wide variety of plants attractive to bees[10, 11].

Since WW2 the intensification of agriculture and the stimulation of production by British governments and the European Commission has increased arable production at the expense of vegetable, fruit and livestock production, particularly in the east of the country. This has been achieved through increased mechanization, larger holdings, and greater use of agrochemicals (fig. 12). Rural areas have been impoverished as an environment for bees, the distribution of beekeepers has changed, and the profitability of beekeeping has been reduced. Particularly severe have been the effects of the use of herbicides, loss of clovers from grassland, changes from deciduous to coniferous woodland and losses of hedgerows and heather moorland.

What are the prospects for the future?

In the UK, four million hectares of agricultural land, more than half, may be surplus to production requirements. To curb overproduction and surpluses of some commodities, the Common Agricultural Policy has, since 1986, been encouraging farmers away from intensive cereal cropping towards land use diversification and lower input farming. Current policy is to take 15% of land out of production through the set-aside policy, under which farmers are compensated for not growing crops. Research is being directed at how to best manage set-aside land, from conservation headlands for game birds, to seed mixes for butterflies[6], but so far little attention has been paid to how the new opportunities could be used for the benefit of bees. Here we explore some of the options for this surplus land that could be of use to bees, and how research, particularly that at Rothamsted, is beginning to address these issues.

The current policy of diversification away from cereal production is expected to bring new crops into agricultural production in the UK, some of which may provide new sources of forage for bees. Among crops being considered are borage, camelina, cosmea, crambe, cuphea, honesty, meadowfoam, niger, vetches, and sea buckthorn. At Rothamsted, an extensive programme for the development of lupins and sunflowers as protein and oil crops is under way[20, 40, 42] and crops are also being developed for industrial uses and for fuels. Their production would not only increase the abundance of food for bees but also the need for bees as pollinators.

The Forestry Commission recently stated Government forest policy to have two main aims: 'the sustainable management of our existing woods and forests', and 'a steady expansion of tree cover to increase the many, diverse benefits that forests provide'. There is currently a target to plant 33 000 ha of woodland

annually plus an additional 12 000 ha under the Farm Woodland Scheme. Special payments are available for small woodland planting, broadleaved planting, the development of community forests in urban fringes and the maintenance of existing woodland. There is also interest in the development of coppice management of willow and poplar for fuel: coppicing allows more light to reach the ground, encouraging growth of useful forage for bees.

Quantitative control of overproduction through the set-aside scheme will benefit bees only if previously intensively cultivated land is converted to bee pasture. Land set aside from agriculture could either be allowed to revert to 'nature' as permanent set-aside, or cultivated as non-rotational set-aside with non-arable crops so as to retain it for possible future agricultural use.

Rothamsted can claim to have the first set-aside experiments in the world. In 1882 about 0.2 ha of a wheat field, Broadbalk Field, was enclosed by a fence, left unharvested and the land not further cultivated. One half of the area remained undisturbed. It is now mature woodland of ash, sycamore and oak with an understorey of hawthorn that is dying out. The ground cover consists of ivy, dog's mercury, violet and blackberry. On the other half, bushes have been hoed out each year to keep the ground open. The bushes that appear are mostly hawthorn, dog-rose, wild plum, blackberry with a few maple and oak. The ground cover consists of coarse grasses, hogweed, agrimony, willow-herb, nettles, knapweed and cow parsley. In 1957, the grubbed section was divided into two parts; half continues to be grubbed of bushes, while the other half was mown several times during the next three years and then grazed by sheep. By 1962 perennial ryegrass and white clover had appeared and were widely distributed. Most species now growing in this abandoned wheat field are of value to bees[33]. A larger area, Geesecroft Wilderness, was set aside in 1886, and plans are now being made to set aside a similar piece of land again to monitor its vegetational changes.

Cultivation of set-aside with non-arable crops could perhaps involve sowing bee forage crops. At Rothamsted phacelia (fig. 13) is being evaluated as a possible candidate for this. Phacelia produces nectar and

FIG. 13. On set-aside land plants like *Phacelia tanacetifolia* can provide bee forage.

Production can also be reduced by using land less intensively, for example by reducing inputs to agricultural production particularly the application of nitrogen fertilizer in the nitrate sensitive areas. This could increase the amount of white clover used in agriculture. White clover fixes its own nitrogen and can reduce input costs and improve livestock performance. However, there is a shortage of seed of the UK-bred cultivars most suited to grassland production, and a programme of research at Rothamsted is investigating pollination as a factor limiting seed production which may help increase use of this very important bee forage plant.

Extensification schemes are expected to lead to a decline in the rate of degeneration of uncultivated areas, more floristically diverse grasslands, reduced stocking levels and decreased use of chemicals. The importance of hedgerows as windbreaks and as wildlife reservoirs and refuges is now more widely appreciated, and may encourage more sympathetic headland and field margin management. The prohibition of straw burning will help preserve bee forage in field boundaries.

The availability of the conservation premium to farmers in the eastern region for the creation of more varied hedgerows, belts of broadleaved trees, meadowland, wildlife fallow and habitat restoration is potentially of considerable value to bees. The woodland option, which allows direct set-aside to woodland or set-aside through the Farm Woodland Scheme, promotes the planting of broadleaved trees rather than conifers and may be beneficial provided that management is sensitive to the needs of bees.

Thus the prospects for the future are good, but beekeepers and conservationists must make their needs known to policy makers and farmers if they are to make the most of the opportunities to increase the bee forage within agroecosystems.

pollen abundantly, has a high flower density and is visited by the honey bee as well as all seven of the mainland ubiquitous bumble bees. Successional sowings can provide flowers from early May to late November[47]. The use of plants such as sainfoin, clovers, mustard and phacelia, or suitable wild flower mixes together with post-flowering cutting, will enhance set-aside for bees: the use of grasses alone will not.

References

The numbers given at the end of references denote entries in *Apicultural Abstracts*.

1. ANON (1947–1981) *Agricultural statistics, England and Wales (for 1945–1979)*. HMSO; London, UK.
2. ANON (1947–1990) *Agricultural statistics, United Kingdom (for 1945–1989)*. HMSO; London, UK.
3. ANON (1955–1992) *Survey of bee health in England and Wales*. HMSO; London, UK.
4. ANON (1981–1993) *Basic horticultural statistics for the UK (for 1980–1992)*. HMSO; London, UK.
5. ANON (1991–1993) *The digest of agricultural census statistics (for 1990–1992)*. HMSO; London, UK.
6. CLARKE, J (ed) (1992) *Set-aside*. British Crop Protection Council; Monograph No. 50.
7. CORBET, S A; WILLIAMS, I H; OSBORNE, J L (1991) Bees and the pollination of crops and wild flowers in the European Community. *Bee World* 72(2): 47–59.   1449/91
8. DEANS, A S C (1957) *Survey of British honey sources*. Bee Research Association; London, UK; 20pp.   195/65
9. FREE, J B (1993) *Insect pollination of crops*. Academic Press; London, UK; 684pp.   1412/93
10. FREE, J B; GENNARD, D; STEVENSON, J H; WILLIAMS, I H (1975) Beneficial insects present on a motorway verge. *Biological Conservation* 8: 61–72.   747/77
11. FREE, J B ; WILLIAMS, I H (1980) The value of white clover *Trifolium repens* L. cultivar S100 on motorway verges to honeybees *Apis mellifera* L. *Biological Conservation* 18(2): 89–92.   1272/82
12. FUSSELL, M; CORBET, S A (1991) Forage for bumble bees and honey bees in farmland: a case study. *Journal of Apicultural Research* 30(2): 87–97.   915/92
13. FUSSELL, M; CORBET, S A (1992) Flower usage by bumble bees: a basis for forage plant management. *Journal of Applied Ecology* 29: 451–465.
14. FUSSELL, M ; CORBET, S A (1992) The nesting places of some British bumble bees. *Journal of Apicultural Research* 31(1): 32–41.   401/93
15. HEEMERT, C; RUIJTER, A DE; EIJNDE, J VAN DEN; STEEN, J VAN DER (1990) Year-round production of bumble bee colonies for crop pollination. *Bee World* 71(2): 54–56.   1093/91
16. HERROD-HEMPSALL, W (1937) *Beekeeping new and old; volume 2*. British Bee Journal; London, UK; 1070pp.
17. HODGES, D (1958) A calender of bee plants. *Bee World* 39(3): 63–70.   337/59
18. HOWES, F N (1945) *Plants and beekeeping*. Faber & Faber; London, UK; 224pp.
19. IUCN (1977) *List of rare, threatened and endemic plants*. International Union for the Conservation of Nature; Kew, UK.
20. LUTMAN, P J W; FITT, B D L; WILLIAMS, I H (1991) Alternative crops. *In* Agriculture and Food Research Council (ed) *AFRC Institute of Arable Crops Research report for 1991*; pp 51–54.
21. MACDONALD, A; ARMSTRONG, H (1989) *Methods for monitoring heather cover*. Nature Conservancy Council; Research and Survey in Nature Conservation No. 27.
22. MANLEY, R O B (1937) *Honey production in the British Isles*. Faber & Faber; London, UK; 328pp.
23. MANLEY, R O B (1946) *Honey farming*. Faber & Faber; London, UK; 293pp.
24. MANLEY, R O B (1948) *Bee-keeping in Britain*. Faber & Faber; London, UK; 439pp.
25. MARKS, M F (1989) *A hundred years of British food and farming: a statistical survey*. Taylor and Francis; London, UK; 275pp.
26. MOWFORTH, M A; SYDES, C (1989) *Moorland management, a literature review*. Nature Conservancy Council; Research and Survey in Nature Conservation No. 25.
27. NATURE CONSERVANCY COUNCIL (1989) *Woods, trees and hedges: a review of changes in the British countryside*. Nature Conservancy: Focus on Nature Conservation No. 22.
28. OSBORNE, J L; WILLIAMS, I H; CORBET, S A (1991) Bees, pollination and habitat change in the European Community. *Bee World* 72(3): 99–116.   341/92
29. O'TOOLE, C; RAW, A (1992) *Bees of the world*. Blandford Press; London, UK; 192pp.   742/92
30. PERRING, F H (1974) Changes in our native vascular plant flora. *In* Hawkesworth, D L (ed) *The changing flora and fauna of Britain*. Academic Press; London, UK; 461pp.
31. PERRING, F H; FARRELL, L (1983) *British red data books: 1. Vascular plants*. World Wildlife Fund & RSNC; Nettleham, UK (2nd edition).
32. RACKHAM, O (1986) *The history of the countryside*. Dent; London, UK.
33. ROTHAMSTED EXPERIMENTAL STATION (1991) *Rothamsted Experimental Station: guide to the classical field experiments*. Lawes Agricultural Trust; Harpenden, UK; 31pp.
34. TORCHIO, P F (1990) Diversification of pollination strategies for U.S. crops. *Environmental Entomology* 19(6): 1649–1656.   698/93
35. TORCHIO, P F (1991) Bees as crop pollinators and the role of solitary species in changing environments. *Acta Horticulturae* 288: 49–61.   1452/91
36. WILLIAMS, E D (1978) *Botanical composition of the Park Grass plots at Rothamsted 1856–1976* (unpublished).
37. WILLIAMS, I H (1974) *Solitary bees in Britain*. Central Association of Beekeepers; Lectures given at the Leamington Spa Conference, October 1973.   728L/74
38. WILLIAMS, I H (1980) Oil-seed rape and beekeeping, particularly in Britain. *Bee World* 61(4): 141–153.   725/81
39. WILLIAMS, I H (1985) *Oilseed rape and beekeeping*. Central Association of Beekeepers; Lecture series.
40. WILLIAMS, I H (1987) The pollination of lupins. *Bee World* 68(1): 10–16.   1400/87
41. WILLIAMS, I H (1988) The pollination of linseed and flax. *Bee World* 69(4): 145–152.   674/89
42. WILLIAMS, I H (1991) *Research into the pollination of alternative crops at Rothamsted*. Central Association of Beekeepers. Lectures given at the Leamington Spa Conference, October 1990.   157/93
43. WILLIAMS, I H (1993) The interdependence of agriculture and apiculture in the European Community. *Proceedings of the EC workshop on bees for pollination, 2–3 March 1992, Brussels, Belgium*; pp 7–18.

44. WILLIAMS, I H (1994) The dependence of crop production within the European Community on pollination by honey bees. *Agricultural Zoology Reviews* (in press).

45. WILLIAMS, I H; COOK, V A (1982) The beekeeping potential of oilseed rape. *British Bee Journal* 110: 68–70.

46. WILLIAMS, I H; CORBET, S A; OSBORNE, J (1991) Beekeeping, wild bees and pollination in the European Community. *Bee World* 72(4): 170–180. 446/92

47. WILLIAMS, I H; CHRISTIAN, D G (1991) Observations on *Phacelia tanacetifolia* Bentham (Hydrophyllaceae) as a food plant for honey bees and bumble bees. *Journal of Apicultural Research* 30(1): 3–12. 523/92

48. WILLIAMS, I H; CARRECK, N; LITTLE, D J (1993) Nectar sources for honey bees and the movement of honey bee colonies for crop pollination and honey production in England. *Bee World* 74(4): 160–175. 230/94

49. WILLIAMS, P H (1986) Environmental change and the distributions of British bumble bees (*Bombus* Latr.). *Bee World* 67(2): 50–61. 39/87

What are the important nectar sources for honey bees

Peter Roberts

Institute of Earth Studies, Llandinam Building, University of Wales, Aberystwyth, SY23 3DB, UK.

Introduction

A study was commissioned by the Food Sciences Division of the UK Ministry of Agriculture, Fisheries and Food (MAFF) into the techniques and application of pollen analysis of honey (melissopalynology) in the UK[1]: the purpose of this study has been to develop the knowledge of the main nectar sources for honey bees in England and Wales. The project was formally initiated in October 1992 and since then over 400 samples have been analysed from across the UK.

This paper considers results obtained from samples studied before the end of September 1993. After briefly outlining the procedures involved in analysis and a short study of oilseed rape samples, the majority of the paper is devoted to consideration of the natural sources of nectar and a detailed look at the variations in honey types found in Wales. I then consider species-specific variation across the study areas as a whole, and the more common and unusual samples from across England. Finally I outline the implications of this study in terms of future developments, and make some recommendations.

Methods and materials

Samples for analysis were obtained through contacts with individual beekeeping groups. The Welsh

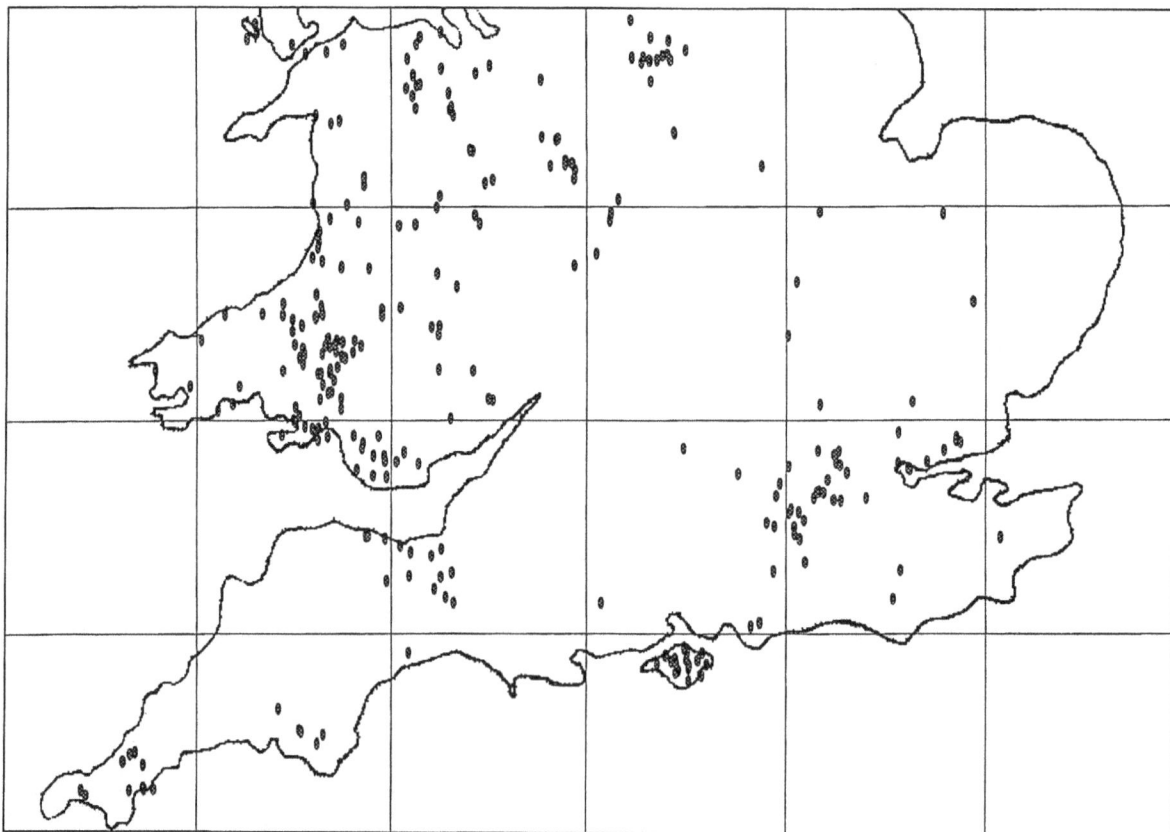

FIG. 1. Distribution of samples across England and Wales.

samples were collected as part of a detailed study of Welsh honey sources which had already been scheduled for 1992. The remainder of the samples are the result of letters to the editors of all county beekeeping association newsletters explaining the project, and attendance at the 1992 National Honey Show in London. Over 2 500 sample jars have been distributed to beekeeping associations outside Wales, of which approximately 350 have been returned.

As a direct result samples have so far been received from over 30 counties in England and Wales, plus Jersey and also a small number from Scotland. As figure 1 illustrates, the majority of these samples are from Wales, the Isle of Wight and the south-east of England, although small clusters are found in the majority of the English counties.

Standardized preparation techniques[1, 2] were used with the sample being acetolysed prior to analysis to increase the detail of the pollen exine. A total of at least 600 pollen grains from each sample were counted to determine the pollen spectrum. Two additional calculations were then undertaken. The first involved the use of published pollen coefficients[4], to correct the pollen percentages to take into account under- and over-represented species. The second used counts of marker *Lycopodium* spores of known concentration to calculate the concentrations of each individual species. Examples of the two sets of percentage data only are presented as separate pollen spectra diagrams for the purpose of this study.

Oilseed rape

This widely grown commercial crop (*Brassica napus*) is currently a very important source of nectar in the UK (fig. 2). Winter rape is of particular relevance as this produces a dominant nectar flow in the late spring to early summer, the honey from which is frequently harvested as a separate crop. This type of oilseed rape honey has been found extensively in the Welsh Marches, Vale of Glamorgan, and the rural areas of central and southern England. Spring sown rape is also an important nectar source in some areas, such as Clwyd, however this flow occurs at a similar time to the main nectar flows of both bramble and white

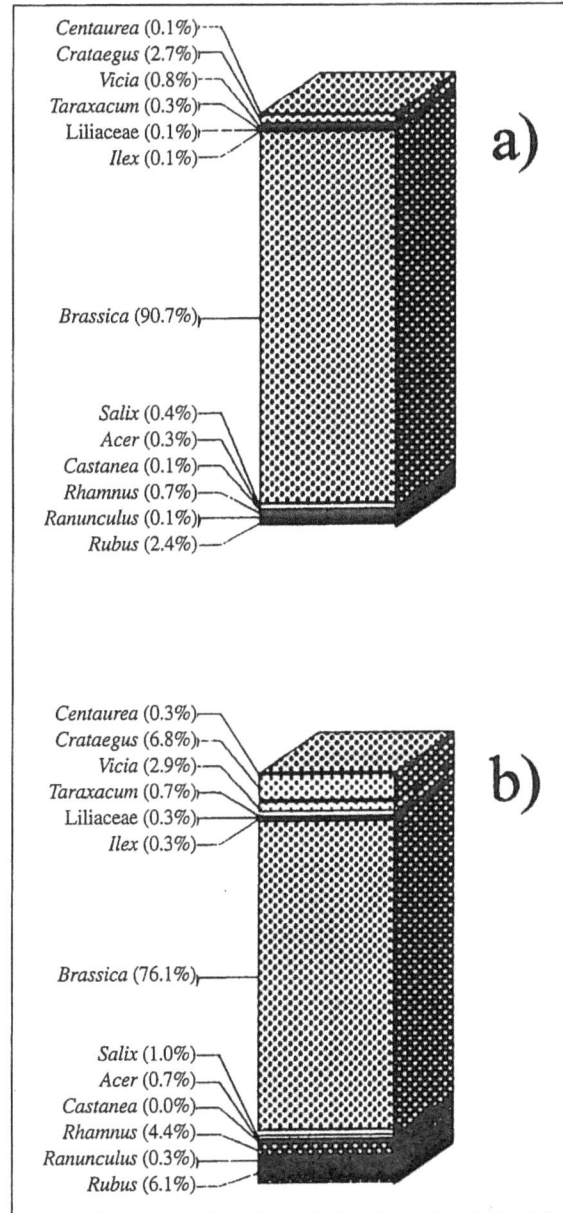

Centaurea (0.1%)
Crataegus (2.7%)
Vicia (0.8%)
Taraxacum (0.3%)
Liliaceae (0.1%)
Ilex (0.1%)
Brassica (90.7%)
Salix (0.4%)
Acer (0.3%)
Castanea (0.1%)
Rhamnus (0.7%)
Ranunculus (0.1%)
Rubus (2.4%)

Centaurea (0.3%)
Crataegus (6.8%)
Vicia (2.9%)
Taraxacum (0.7%)
Liliaceae (0.3%)
Ilex (0.3%)
Brassica (76.1%)
Salix (1.0%)
Acer (0.7%)
Castanea (0.0%)
Rhamnus (4.4%)
Ranunculus (0.3%)
Rubus (6.1%)

FIG. 2. Pollen diagrams for oilseed rape sample PRC/F299 (Isle of Wight): (a) pollen percentages; (b) calculated nectar equivalents.

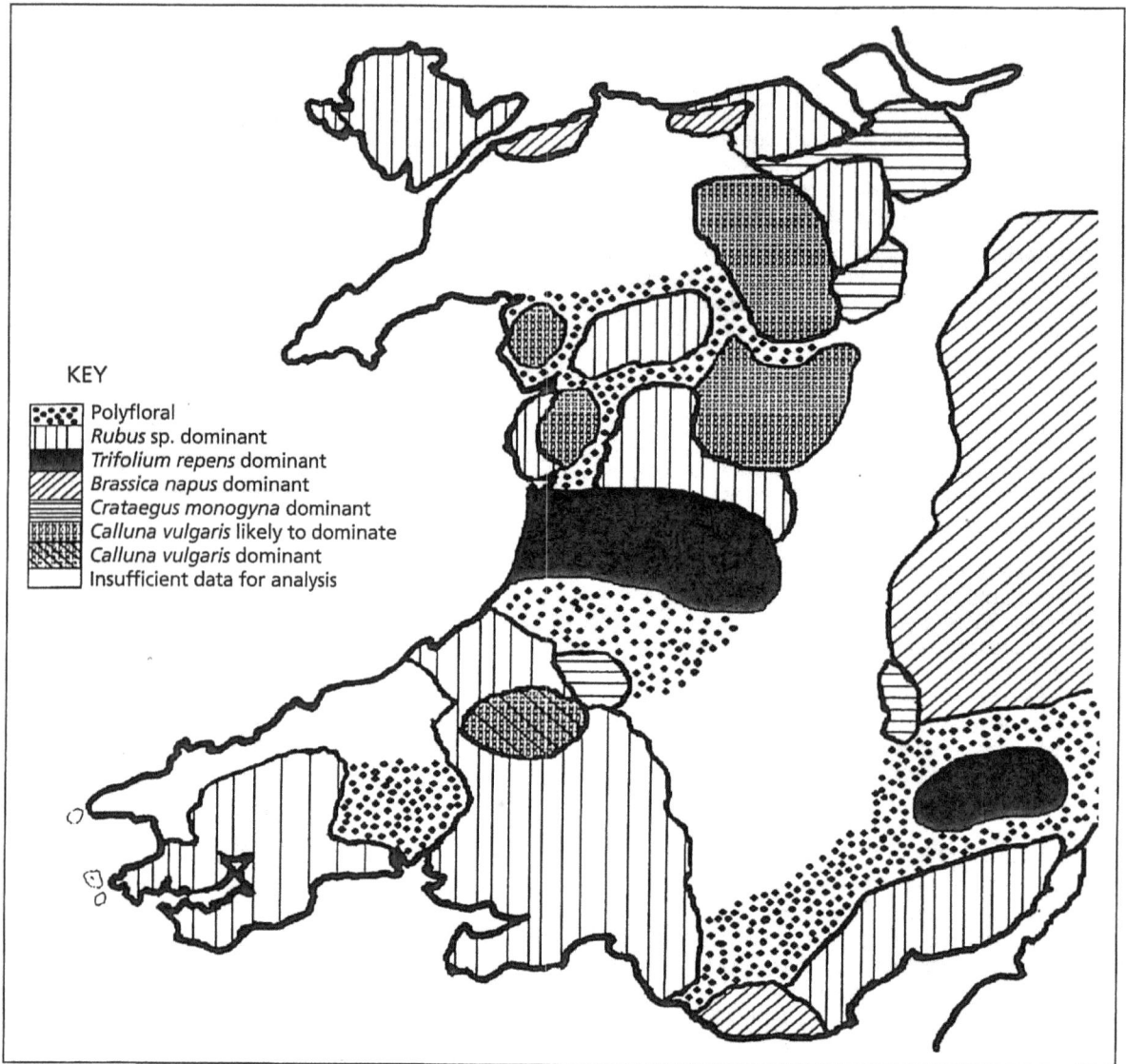

KEY

- Polyfloral
- *Rubus* sp. dominant
- *Trifolium repens* dominant
- *Brassica napus* dominant
- *Crataegus monogyna* dominant
- *Calluna vulgaris* likely to dominate
- *Calluna vulgaris* dominant
- Insufficient data for analysis

FIG. 3. Principal nectar and honey sources present in Wales: a preliminary distribution map.

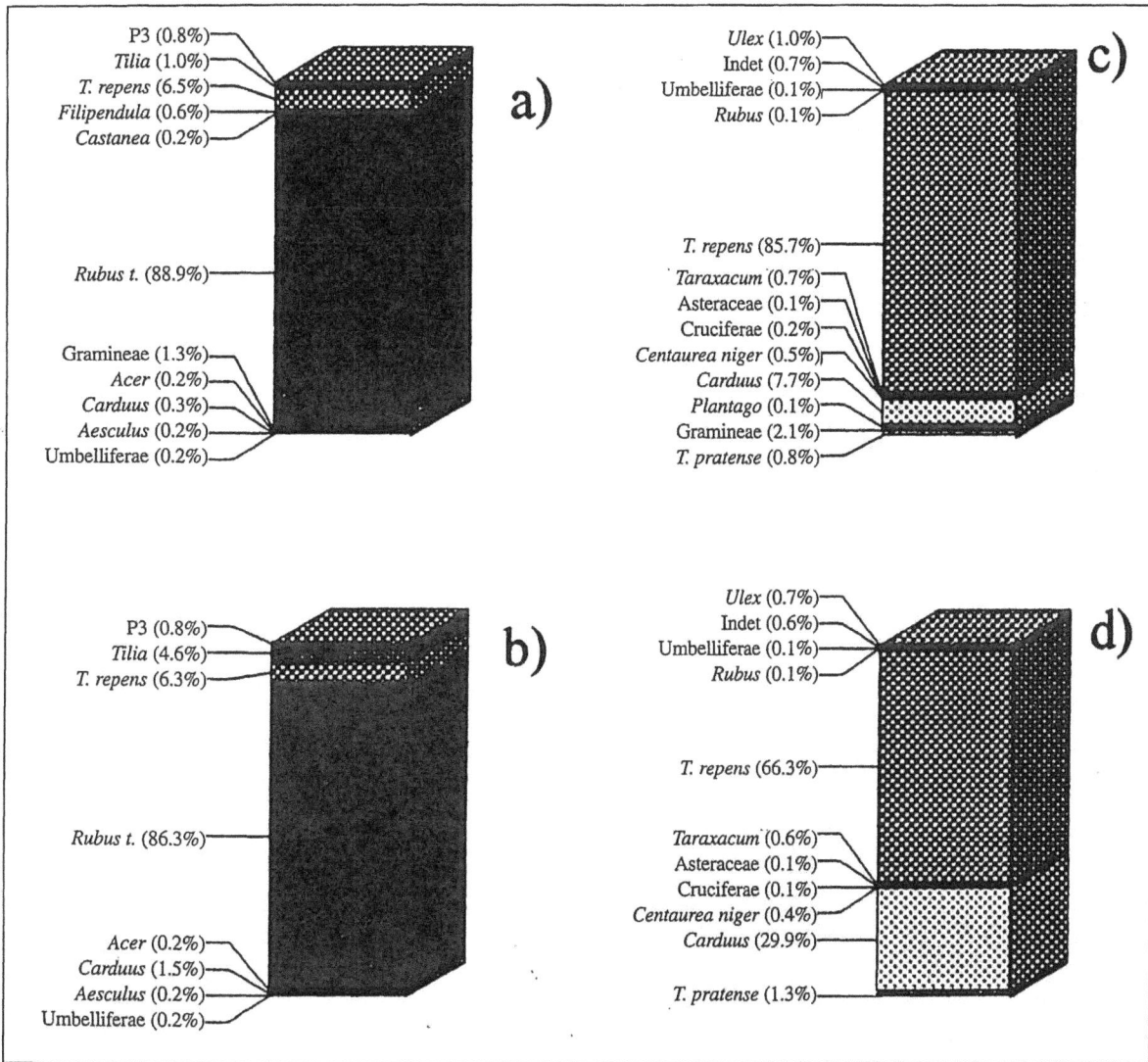

FIG. 4. Pollen diagrams for bramble sample PRC/F173 (Dyfed): (a) pollen percentages; (b) calculated nectar equivalents and white clover sample PRC/F72 (Dyfed); (c) pollen percentages; (d) calculated nectar equivalents.

clover which ensures that the honey produced at that time is more multifloral in nature.

Wales

With the majority of the initial research concentrated on Wales the results obtained there offer the greatest detail to date. At the same time many of these samples can be considered close to the UK's traditional honey sources, since in large sections of Wales agriculture has changed little over the last 50 years. While this research is not yet complete it has been possible to produce an interim map (fig. 3), illustrating the variation in nectar sources found to date at Welsh sample sites. Now in its third draft, the information available covers approximately 66% of the land area of Wales, although a more detailed redraft is expected to be completed in the next couple of months as results from new samples on the Lleyn peninsula and large parts of Gwent, Powys and Pembrokeshire become available.

To date five dominant unifloral honey types have been isolated in Welsh samples: bramble (*Rubus*), white clover (*Trifolium repens*), hawthorn (*Crataegus monogyna*), heather (*Calluna vulgaris*) and oilseed rape. In addition there are a large number of marginal areas where multifloral samples containing these and other species occur. A typical pollen diagram from each of these types is considered as they are reviewed in turn.

FIG. 5. White clover: a traditionally important source of nectar.

Bramble

This type (*Rubus*) is dominant in the south Wales valleys extending up into Pembrokeshire, Carmarthenshire and south Cardiganshire. It also occurs in valley sites throughout much of north Wales and Anglesey. These samples have usually been harvested from comb that has been in the hive for the full season and is extracted at the end of either July or August. While this can be very pure with up to 92% of pollen in a sample originating from *Rubus*, associated species include all the main species found in Wales with additional nectar from rosebay willow-herb (*Epilobium angustifolium*), thistle (*Carduus*) type and field maple (*Acer campestre*) (figs 4a and 4b). In terms of area covered the variants of this source are the most dominant in Wales.

White clover

The traditional white clover, *Trifolium repens* (fig. 5), honey is rarer and has to date been found concentrated in two areas: the first inland from Aberystwyth extending to Builth Wells, and the second in northern Gwent. The honey is often less pure than that from bramble samples, and is commonly found with large amounts of thistle type pollen and also bramble, Ranunculaceae and occasionally red clover (*T. pratense*). Unusually for Wales these samples have very low percentages of hawthorn, principally because the main season of honey production is later in the year (figs 4c and 4d).

Hawthorn

This type (*Crataegus monogyna*) is rarer and is directly related not only to availability of the species but also beekeeping practices. An early flowering and often unreliable crop, hawthorn is invariably finished before beekeepers place the year's frames on the hive, and as a consequence the very small areas where this type has been found as a pure sample are likely to be unrepresentative of the actual potential of the species. Where a sample has been collected hawthorn pollen accounts for over 50% of the sample (figs 6a and 6b), with associated pollens often including willow (*Salix*) and field maple (*Acer campestre*): this

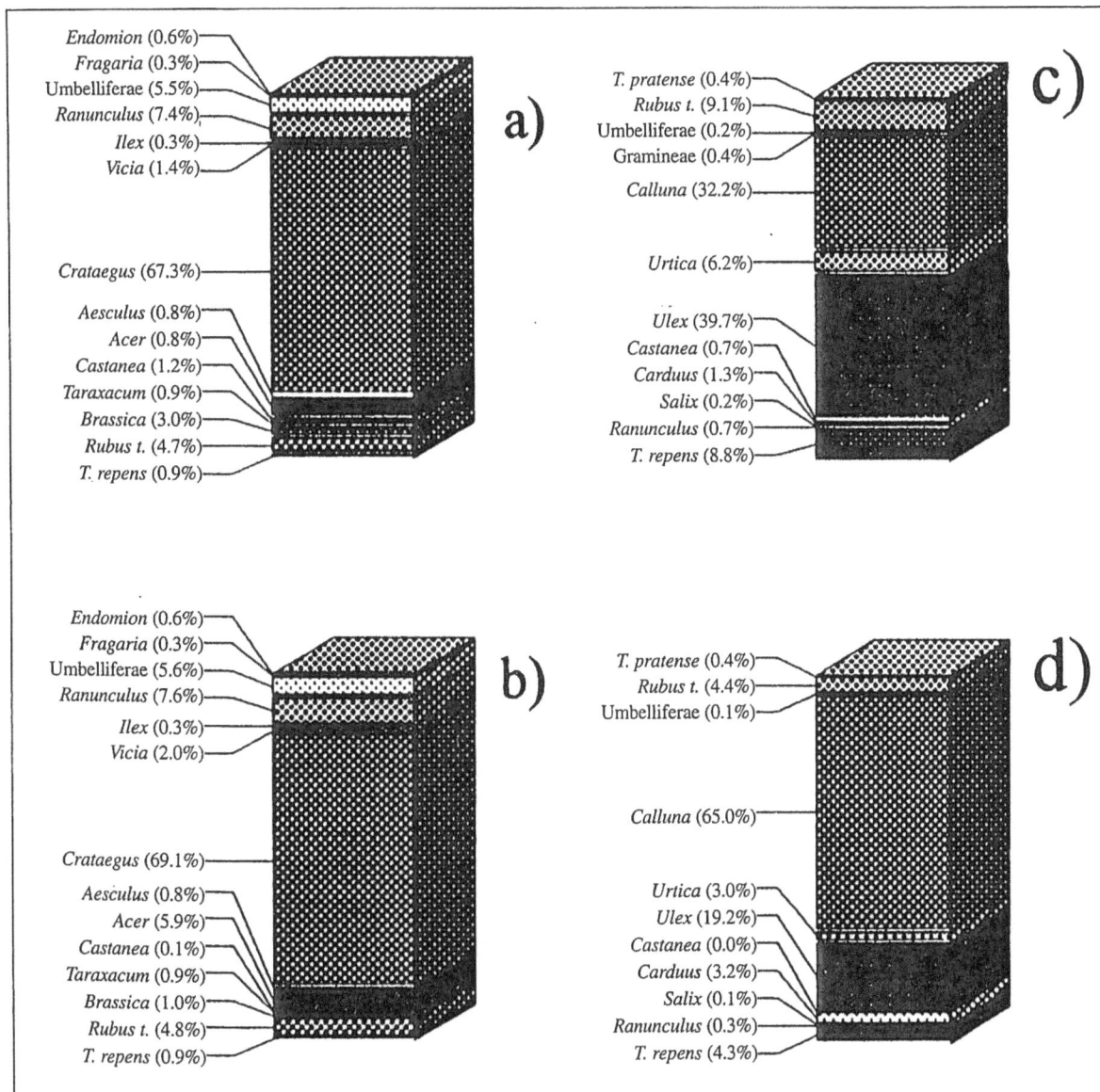

FIG. 6. Pollen diagrams for hawthorn sample PRC/F126 (Dyfed): (a) pollen percentages; (b) calculated nectar equivalents and Heather sample PRC/F68 (Dyfed); (c) pollen percentages; (d) calculated nectar equivalents.

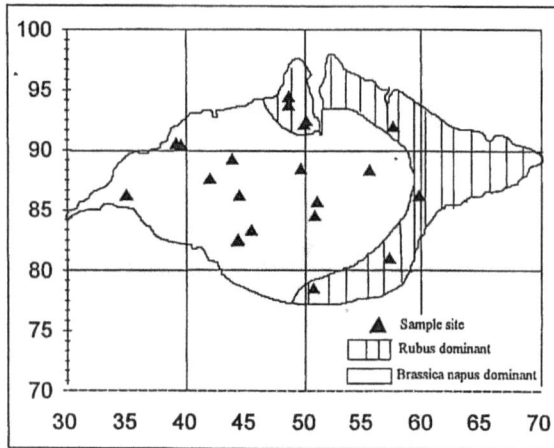

FIG. 7. Variation in honey type across the Isle of Wight.

central Cardiganshire), or else they comprise honey extracted that is a combination of two distinct honey flows (in the Welsh Marches). Within the definition multifloral there is no overall consistency in the honey type, with local variations important.

The nectar sources used by honey bees in Wales have been extensively documented, and this study shows a distinctive pattern of natural honey flows. This situation reflects the fact that large areas of Wales are used as grazing land for sheep and cattle, and in these areas the traditional method of field enclosure appears to be vital in determining the type of honey produced. In the upland areas where walls dominate, the honey is more likely to be from white clover, while in the lowland areas and where cattle are more important than sheep hedgerows dominate, so the honey produced is dominated by bramble, hawthorn and other hedgerow species.

England

The distribution of samples from England is also clustered, however the overlap of these clusters is very small so there is still considerable research required before true regional variations can be mapped.

Isle of Wight

The exception to this statement is the Isle of Wight, from where 23 samples have been received. Oilseed rape and brambles are the most important nectar sources, accounting for an average of over 30% each of the nectar. Of the remaining taxa only white clover, field bean (*Vicia faba*) and hawthorn exceed an average of 3% each in the samples. The same is true of the occurrence of pollen in the samples, with oilseed rape and brambles present in all 23 samples. White clover, buttercup family (Ranunculaceae), willow (*Salix* spp.), Umbelliferae, field bean and hawthorn are also present in over 75% of the samples.

These figures, however, do not represent a uniform distribution of nectar sources, but show the island's distinct split into two main sectors (fig. 7). The first of these, predominantly to the west, is a farming area. Here there are two main honey flows: the first

reinforces the idea of an early season nectar collection. To date samples have been collected in the Lampeter area, northern Clwyd and around Brecon.

Heather

The final natural source defined by this study, heather (*Calluna vulgaris*), has so far been confirmed only in a very few areas of Wales with samples having been received from the Preselis and north of Machynlleth: the samples expected from large parts of southern Snowdonia have not yet appeared. Those samples we have seen are from previously used comb, and thus contaminated by other honey flows. However, it appears that gorse (*Ulex europaeus*) and rosebay willow-herb are important secondary sources during heather honey production (figs 6c and 6d).

Other honey types

As previously noted, samples of oilseed rape honey are found around the English border, in the Vale of Glamorgan and also in isolated areas of Pembrokeshire and the north Wales coast. Wales also produces a very varied collection of multifloral honeys: usually these are the results of hives falling in between two areas of distinctive sources (such as in

TABLE 1. Species noted in one or more sample during the current study.

Aceraceae
Acer campestre
Acer pseudoplatanus

Aquifoliaceae
Ilex aquifolium

Araliaceae
Hedera helix

Balsaminaceae
Impatiens glandulifera

Betulaceae
Betula sp.

Boraginaceae
Borago officinalis
Echium vulgare
Lithospermum
   purpurocaeruleum
Symphytum officinale
Myosotis arvensis

Buddlejaceae
Buddleja davidii

Campanulaceae
Campanula dioica
Jasione montana

Caprifoliaceae
Lonicera sp.
Sambucus nigra

Caryophyllaceae
Silene dioica

Chenopodiaceae
Chenopodium alba

Cistaceae
Cistaceae undiferentiated
Helianthemum sp.

Compositae
Anthemis type
Aster type
Bidens type
Carduus type
including
   Carduus acanthoides
   Cirsium sp.
Centaurea type

including
   Centaurea nigra
   Centaurea scabiosa
Helianthus annuus
Taraxacum type
including
   Cichorium intybus
   Taraxacum officinale
   Crepis sp.
   Soncus sp.

Convolvulaceae
Calystegia sepium
Convolvulus sp.

Corylaceae
Corylus sp.

Cruciferae
Brassica napus
Sinapis alba

Cucurbitaceae
Bryonia dioica

Cupressaceae
Juniperus communis

Dipsacaceae
Dipsacus pilosus
Knautia arvensis

Ericaceae
Calluna vulgaris
Erica type
including
   Erica cinerea
   Erica tetralix
Rhododendron ponticum
Vaccinium

Fagaceae
Castanea sativa
Fagus sylvatica
Quercus sp.

Gossulariaceae
Ribes sp.

Gramineae
Zea mays

Hippocastinaceae
Aesculus hippocastanum

Hydrophyllaceae
Phacelia tanacetifolia

Labiateae
Lycopus europaeus
Mentha type
Rosmarinus officinalis

Liliaceae
Endymion non-scriptus
Iris sp.

Linaceae
Linum sp.

Lythraceae
Lythrum album

Malvaceae
Malva sylvestris

Myricaceae
Myrica gale

Myrtaceae
Eucalyptus sp.

Nymphaeaceae
Nuphar luteum

Oleaceae
Ligustrum vulgare

Onagraceae
Epilobium angustifolium
Fuchsia sp.

Papilionaceae
Onobrychis viciifolia
Trifolium repens
Trifolium pratense
Ulex europaeus
Vicia faba

Pinaceae
Pinus sp.
Picea sp.

Plantaginaceae
Plantago type

Polygonaceae
Polygonum sp.
Rumex acetosella type

Ranunculaceae
Helleborus niger

Ranunculus type
including
   Clematis alba
   Ranunculus ficaria

Rhamnaceae
Frangula alnus
Rhamnus catharticus

Rosaceae
Crataegeus monogyna
Filipendula ulmaria
Fragaria sp.
Malus sylvestris
Prunus avium
Pyrus sp.
Rubus type
including
   Rubus fruticosus
   Rubus ideaus
   Cotoneaster sp.
Sorbus sp.

Rubiaceae
Galium album

Salicaceae
Salix sp.

Saxifragaceae
Saxifraga sp.

Tiliaceae
Tilia sp.

Ulmaceae
Ulmus sp.

Umbelliferae
Umbelliferae type
including
   Daucus carota
   Heracleum sphondylium

Urticaceae
Urtica dioica

Valerianaceae
Valeriana sp.

Violaceae
Viola sp.

Vitaceae
Vitis sp.

dominated by pure oilseed rape honey, and the second retaining some of these characteristics but also including other species such as field bean, buckthorn (*Rhamnus catharticus*) and bramble.

The second region is mainly urban in its composition and here the honey comes from a single flow that is dominated by bramble type pollen with no main secondary sources occurring in all samples: although privet (*Ligustrum vulgare*), field bean and hawthorn occur in most.

Southern England

This pattern is repeated across large parts of southern England, with the rural areas producing two or more distinct and often unifloral honey flows and the urban areas producing honey from a far greater diversity of nectar sources which is often harvested only at the end of the season.

Variations in honey sources

Results from the study so far allow me to make some generalizations. Firstly there is a considerable variation in honey sources. Over 100 pollen types have been identified in the samples processed to date (table 1), with many occurring in only a handful of samples. However, in a total dataset of approximately 400 samples (including Wales and Scotland) the 25 species shown in figure 8 were the most common in their occurrence. While not all of these are believed to be nectar producers (and this graph is based only on occurrence of the species), it does illustrate the most common plants that bees visit at some stage during the year.

Taking a closer look at the diagram, many of the lines indicate not a single species but a pollen type shared by several. Umbelliferae, for example, encompasses five different morphotypes ranging from *Daucus carota* (wild carrot) to *Heracleum sphondylium* (hogweed).

The obvious and considerably large variation in pollen sources leads to consideration of the most important species in turn, and involves looking at a fixed percentage value for each species and examining the number of samples that exceed this value on a county by county basis. This process has enabled me to produce table 2, which starts to illustrate the relative importance of different species in various parts of the country. At the moment this part of the research is at a very early stage: the results presented are likely to alter considerably after further analyses are completed, because of the very small datasets used to date.

Atypical results

Finally in this section I want to consider a few oddities. So far I have concentrated on the species that dominate on a county-wide basis, however, at individual sites differences do occur. On the Isle of Wight, for example, a single sample of (*Phacelia tanacetifolia*) honey was produced by careful use of set-aside land. The same experiment has been repeated in Wales, and the species is a common component of samples from Jersey. The fact that this plant has been successfully cultivated in the UK indicates that in the future it may provide a significant nectar source not only for honey bees but also bumble bees, as it is noted as an important food source[3].

In central England, around Reading, a series of samples was received that had higher than usual levels of lime (*Tilia*), and in central London a pollen type which may be *Ailanthus* is very common. In suburban Middlesex a single sample had over 20% of *Eucalyptus* pollen present, while in the home counties several samples had been produced which had borage (*Borago officinalis*) as the main nectar source. These are at the moment unusual sources, but they do illustrate that there is a large potential for variation in nectar sources in the UK which is as yet unexploited.

Implications

The application of this work remains to be fully exploited, and its true potential will only become apparent when more samples have been processed to give wider coverage. However, there are a number of important results that have already been obtained that are of immediate relevance to beekeeping in the 1990s and beyond.

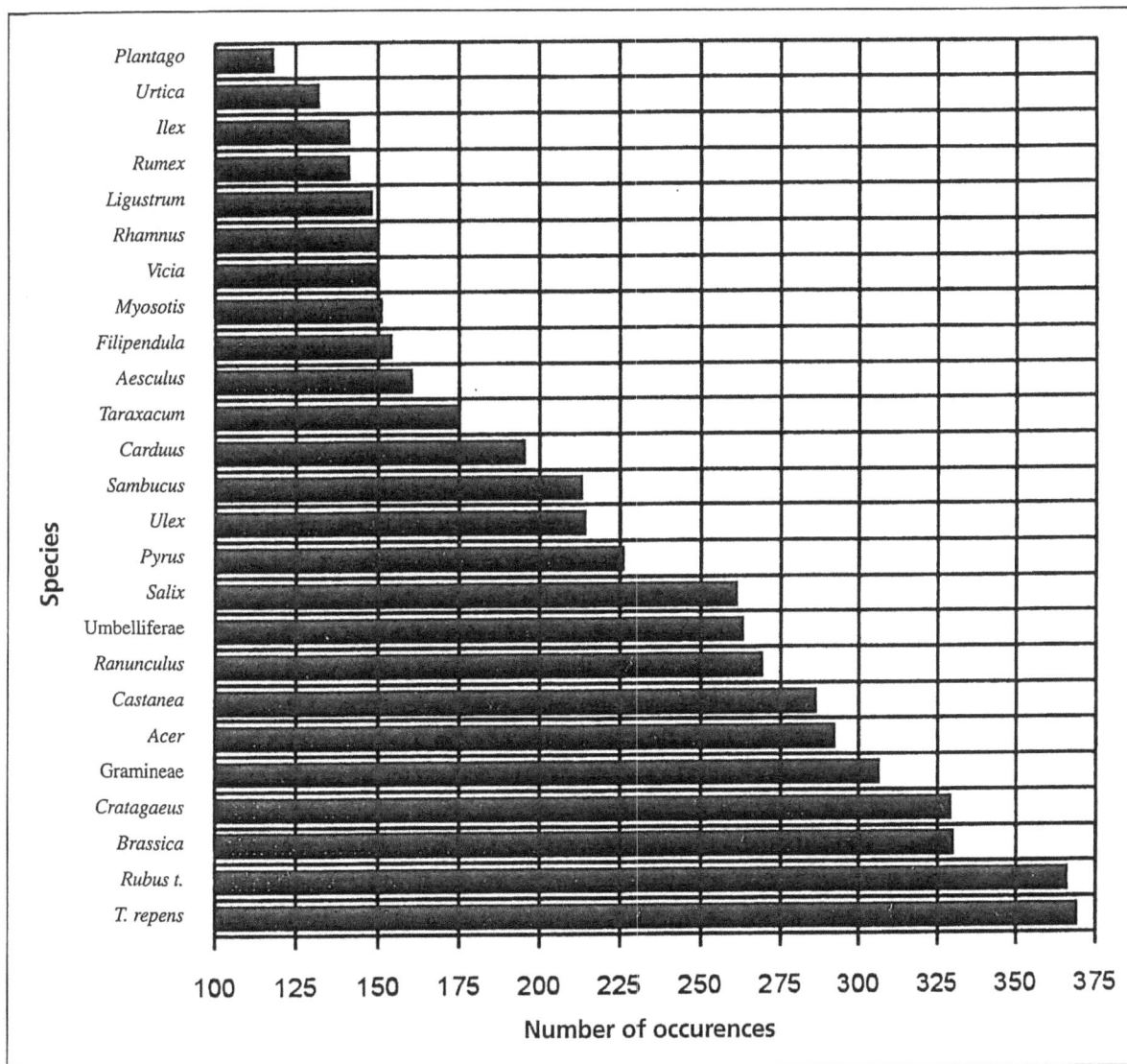

FIG. 8. Number of occurrences of the 25 most common pollen types.

TABLE 2. Honey types from the UK, based on the percentage of samples from any single county exceeding a specified pollen percentage.

Honey type and pollen source	Percentage of samples in which pollen source concentration exceeds given figure			
	> 20%	> 40%	> 60%	> 80%
Bramble *Rubus* > 25%	W Midlands, Surrey, Staffs, Isle of Wight, Essex	Somerset, Powys, Middx, Devon, Derby	M Glam, Dyfed, Clwyd	
Bramble *Rubus* > 50%	Powys, Derby, Cornwall, Clwyd	Gwynedd, Gwent	Dyfed	W Glam
Bramble *Rubus* > 75%	M Glam, Dyfed, Gwynedd	W Glam		
White clover *T. repens* > 20%	M Glam, Gwynedd, Dyfed		Cornwall	
White clover *T. repens* > 50%	Cornwall			
Oilseed rape *Brassica* > 25%	Somerset, Devon, Derby	Powys		
Oilseed rape *Brassica* > 50%	Cornwall, Derby, Powys	Staffs, Essex, Herts, Isle of Wight, W Midlands		
Oilseed rape *Brassica* > 75%	Essex, Isle of Wight, Staffs			
Hawthorn *Crataegus* > 20%	Devon			
Hawthorn *Crataegus* > 10%	Powys, Clwyd, Dyfed, Middx, Jersey, Gwent			
Field bean *Vicia* > 5%	Staffs	Somerset		
Carrot family Umbelliferae > 5%		Jersey		
Maple *Acer* > 5%	Jersey, Clwyd, Gwynedd			
Sweet chestnut *Castanea* > 25%		Surrey, Jersey		

Abbreviations: Herts = Hertfordshire; M Glam = Mid Glamorgan; Middx = Middlesex; Staffs = Staffordshire; W Glam = West Glamorgan; W Midlands = West Midlands

Firstly there is the government's policy of encouraging the preservation and replanting of hedgerows. These could provide a vital source of nectar if the correct species are planted. Based on current research I would encourage the use of the following natural nectar sources: hawthorn, field maple (*Acer pseudoplatanus*) and holly (*Ilex aquifolium*), although privet, and in southern England buckthorn, are also valid choices.

For the longer term there is a proposal for large areas of new broadleaf forests: for the beekeeper a balance of wild cherry (*Prunus avium*) and related species combined with sweet chestnut (*Castanea sativa*), horse chesnut (*Aesculus hippocastanum*), and in the south and south-west lime, would offer large extents of forage.

The other option for set-aside is to return to a traditional pasture sward, and here the traditional white and red clover come to the fore. However, if these are the only available species planted then a spell of poor weather in late June or early July can can be extremely detrimental to the production of honey. Other species should be available to provide a useful supplement: these could include phacelia, thistles, dandelion (*Taraxacum officinale*) and field bean.

Lastly there is the question of use of the uplands: research shows that there is currently a contraction of heather moorland and this may be reflected in the relatively few samples of heather honey received. This problem, due in the main to bracken encroachment, must be addressed, however the expansion of gorse offers potential for exploitation by beekeepers.

The other major opportunity for beekeepers is in the Forestry Commission areas, many of which are being clear felled at the moment. The result of this is a large area of open land which is rapidly colonized by almost pure stands of rosebay willow-herb which can be successfully exploited by bees.

Acknowledgement

This project was undertaken while I was in receipt of MAFF project grant N2704.

References

The numbers given at the end of references denote entries in *Apicultural Abstracts*.

1. LOUVEAUX, J; MAURIZIO, A; VORWOHL, G (1978) Methods of melissopalynology. *Bee World* 59(4): 139–157.   1078/79
2. ROBERTS, P D (1993) *Verification of floral content of honey using melissopalynology*. MAFF; London, UK (unpublished).
3. PATTEN, K D; SHANKS, C H; MAYER, D F (1993) Evaluation of herbaceous plants for attractiveness to bumble bees for use near cranberry farms. *Journal of Apicultural Research* 32(2): 73–79.   229/94
4. SAWYER, R (1988) *Honey identification*. Cardiff Academic Press; Cardiff, UK; 115pp.   1284/89

Farmland as a habitat for bumble bees

Sarah A Corbet[1]; Naomi M Saville[1]; Juliet L Osborne[2]

[1]Department of zoology, University of Cambridge, Downing Street, Cambridge, CB2 3EJ,UK.

[2]Environment and Sustainability Institute, University of Exeter, Penryn Campus, Penryn, Cornwall, TR10 9FE, UK.

Bumble bees as pollinators

Bumble bees provide an essential pollination service for some crops and wild flowers. In some cases this is a background pollination service which can be supplemented by bringing in honey bee hives when the crop is flowering. In other cases bumble bees are the only effective pollinators.

For example, bumble bees visit flowers at times when honey bees are inactive in cold weather, and bumble bees can work some flowers in which the reward is inaccessible to honey bees. On deep flowers like red clover and field bean, only bumble bees with long tongues can reach the nectar: a honey bee's tongue is too short[3, 5]. On borage, a buzz-pollinated flower, only bumble bees can vibrate the anthers at a high enough frequency to release clouds of pollen, as honey bees cannot produce this high-pitched buzz[4, 19]. If we lost our bumble bees, we would lose some wild flower species that depend on them for pollination, and some seed crops would be abandoned by farmers because of low or unreliable yield[20].

Bumble bees have declined

The diary of a naturalist recorded 10 species as common in 1930 in this part of Cambridge[33], where we now have only six common species. Information from national mapping schemes and collections shows that the decline is general in Britain and Europe. Bumble bee species have been lost from parts of the United Kingdom, particularly in eastern central England where there were 14 species before 1960 but there are now only seven[36]. Over the same period bumble bee species have also gone extinct locally in regions of France, Belgium and Germany[23, 25]. The decline has been attributed to changes in agricultural land use eroding the patches of established vegetation that provide resources for bumble bees.

Habitat requirements

The resources that bumble bees need change through the annual cycle[24]. Bumble bees are social insects. In mid-summer, a nest may contain perhaps 50–200 workers. In late summer new queens and males emerge from the nest. The males fly repeatedly along patrolling routes, stopping at intervals to leave a scent mark[13]. Queens are thought to visit and mate here. The mated queen will overwinter underground, and the other bees die. In spring, the queen emerges, feeds on pollen and nectar, and then seeks a place to establish a nest in which she will lay her first batch of eggs. Unlike a honey bee queen, she starts the season as a single parent with no workers to help her, and she forages alone for nectar and pollen to supply herself and her first brood. When worker bees develop from her first batch of eggs, they help to provision the growing colony, and eventually the queen stops foraging and stays in the nest. If a habitat is to support thriving bumble bee populations it must therefore provide patrolling sites for males, overwintering sites for queens and, most importantly, nesting places and a seasonal succession of forage sources.

In early spring, when the queen must forage alone for the pollen and nectar she needs to mature her eggs and provision her first brood, the weather is often cold and dull. In the brief sunny spells, she must forage very effectively if she is to succeed in establishing a colony. The few plant species that produce nectar-rich flowers at this time of year are therefore crucial. Pussy willow is important for queens of the two-banded white tails, and white dead nettle, *Lamium album*, and flowering currant, *Ribes sanguineum*, are among the flowers visited by queens of the longer-tongued browns and three-banded white tails. Gorse provides pollen early in the year[15].

Nesting places

Bumble bees generally nest at the soil surface or underground. The detailed choice of nesting place varies from one species to another. For example, a recent public survey showed that the short-tongued two-banded white tails, *Bombus terrestris* and *B. lucorum*, usually nest underground, often in the abandoned burrows of small mammals; the long-tongued brown bumble bee, *B. pascuorum*, makes a nest of moss on the soil surface in rough grassland; the warmth-loving black-bodied red tail, *B. lapidarius*, often nests beside rocks and masonry that heat up in the sun; and the little banded red tail, *B. pratorum*, is an opportunist, sometimes choosing unexpected above-ground sites such as bird boxes or abandoned waistcoat pockets[14]. Since bumble bees often nest in or near the disused nests of small mammals, birds or previous generations of bumble bees, disturbance of the soil surface by ploughing or herbicide treatment may destroy their nesting places. Generally, bumble bees need undisturbed rough grassland in which to nest.

Forage

Undisturbed rough grassland also provides forage for bumble bees. Our knowledge of the forage plants used by British bumble bees comes largely from a public survey organized in conjunction with WATCH, the junior wing of the Royal Society for Nature Conservation[15, 34], and a study of bumble bee forage usage in farmland and farm woodlands in Cambridgeshire[10, 29]. As in the case of nesting places, bumble bee species differ from one another in their choice of forage plants. A bumble bee colony stores very little honey and cannot survive for more than a day or two without foraging; it needs a continuous succession of suitable flowers from spring to autumn. In general, each bumble bee species visits flowers of a characteristic type. A bumble bee species that will pollinate a particular crop needs other flowers of a similar type to support it for the rest of the year.

FIG. 1. *Bombus pascuorum* taking nectar from marsh woundwort, *Stachys palustris*. (photo: S A Corbet)

The survey has helped us to dèfine the characteristic flower types used by the major bumble bee species. All bumble bees are large insects with a high energy requirement, both for flight and for the muscular effort required to generate the heat that warms the nest and allows them to forage in cold weather. Since their costs are high, they need to collect a large amount of nectar sugar from each flower to make a profit. Bumble bee flowers are typically large flowers that offer abundant nectar (fig. 1).

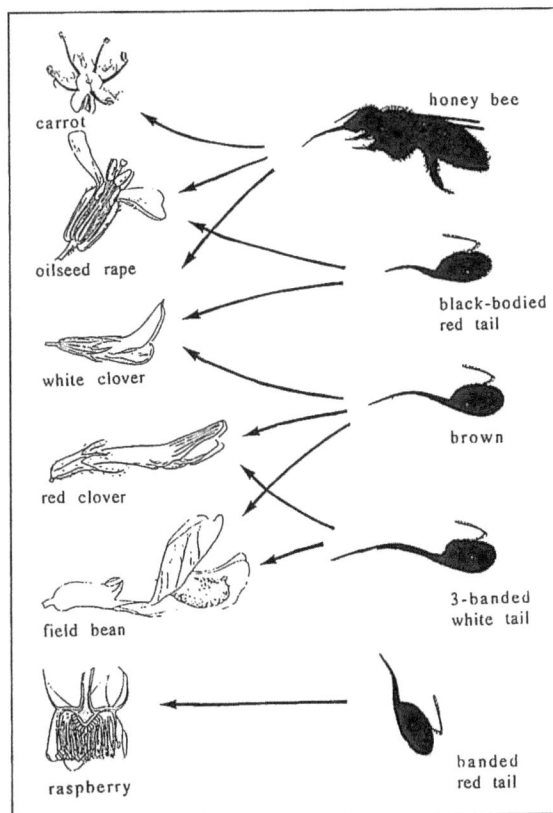

FIG. 2. Flower selection by nectar-foraging bees depends on the relationship between bee tongue length and flower form. Heads are shown for the following bumble bees: black-bodied red tail (*Bombus lapidarius*); brown (*B. pascuorum*); three-banded white tail (*B. hortorum*); and banded red tail (*B. pratorum*), which often visits pendulous flowers.

Crop	Bumble bee colour group				
	Two-banded white tail	Banded red tail	Black-bodied red tail	Brown	Three-banded white tail
Oilseed rape	+		+		
Raspberry	+	+	+	+	
Borage	+	+	+	+	
Birdsfoot trefoil			+		
White clover			+	+	
Field bean	rob			+	+
Red clover	rob			+	+

TABLE 1. Some crops and the bumble bee colour groups that readily visit them[7, 15, 19, 28].

All bumble bee species are alike in this requirement for nectar-rich flowers, but the species differ from one another in the types of flowers they choose. An important feature is the depth of the flower. Bumble bee species with long tongues, such as the three-banded white tail and the brown bumble bee, visit deep flowers, whereas those with shorter tongues, such as the two-banded white tails, are more like honey bees in both tongue length and flower choice, concentrating on shallow or open flowers with more accessible nectar (see fig. 2).

Long-tongued species are particularly valuable pollinators, because they cannot be replaced by honey bees on deep-flowered crops. A good farm should have large numbers of the brown bumble bee, and the even longer-tongued three-banded white tail should also be present although it is rarely as abundant as the brown. Both of these species pollinate field bean and red clover, and the brown visits many other species including raspberry, tomato and white clover (table 1).

The shorter-tongued species, the two-banded white tails *B. terrestris* and *B. lucorum*, have tongues about the same length as honey bees. They frequently collect pollen on flowers that yield no nectar, and they are exploited commercially for the pollination of such flowers, for example glasshouse tomatoes. The queens emerge early and their ability to forage in cold weather makes them useful pollinators of apple blossom. But these short-tongued bumble bees are sometimes robbers. When they cannot reach the nectar in deep flowers by probing from the front of the flower, they sometimes bite a hole in the corolla and take nectar without pollinating. Honey bees and other bumble bees sometimes re-use these holes, for example on red clover or field bean[9, 12]. When recording which bees are visiting a crop one should look carefully to distinguish robbers from visitors that might be pollinating the flowers.

Bumble bees usually fly from flower to flower, and they choose plants in which the individual flowers are large enough to pay the cost of these flights; but the black-bodied red tail *B. lapidarius* often lowers its costs by scurrying on foot between clustered flowers, some of which may be individually small. It favours members of the daisy family, including thistles and knapweeds, and legumes with clustered flowers such as white clover and birdsfoot trefoil[15, 24].

An unexpected finding from the national survey was a general preference of bumble bees for perennials rather than annuals. Bumble bees rarely visited annual plants, perhaps because most annuals are small plants with small flowers that yield little nectar. Most bumble bee visits were made to biennials and perennials, which are generally larger, longer-lived plants with larger flowers and more nectar[15].

The national survey was conducted by inexperienced observers and the records were numerous, covering wide geographical area, but necessarily simple. More detailed investigations in local farmland confirmed the bumble bees' preference for perennials ove annuals, and added further information. They showed that biennials are visited even more than longer-lived perennials, and that the preference for biennials and perennials is shown by honey bees and butterflies as well as bumble bees[8, 10, 29].

Managing habitats for bumble bees

These findings have implications for land management. A few perennial weeds may survive cultivation as vegetative fragments in the soil, but apart from these, annuals are typically the only plants that flower in the first year after ploughing or herbicide treatment. Since bumble bees and honey bees rarely visit these, preferring biennials and perennials that take longer to establish, vegetation that has been undisturbed for several years provides better forage than plants that are early colonists of bare soil in cultivated fields or road verges. If the sward continues undisturbed, with occasional mowing to prevent scrub invasion, floristic diversity will increase progressively, helping to provide a range of flowers for all species in all seasons.

We stress the high value, and irreplaceability, of established perennial herbaceous vegetation. Very long-term set-aside can eventually supplement it to some extent, but even small fragments of vegetation that are decades or centuries old are worth treasuring. They form a precious refuge for wild bees and other wildlife, and a source of colonization for regeneration elsewhere on the farm. The value of a set-aside policy for bumble bees can be measured in terms of the conservation of these precious fragments. Any policy that results, intentionally or inadvertently, in the destruction of perennial herbaceous vegetation, or prevents its establishment, will diminish the capacity for the farming landscape to support bumble bees, threatening the survival of the few species that still remain. A policy that safeguards existing perennial vegetation, and helps to extend it, will help to reduce the risk of further losses of wild pollinators.

Areas of established perennial herbaceous vegetation are essential to provide nesting places and a seasonal succession of forage for bumble bees and honey bees

FIG. 3. Unsprayed edge and plantation next to field bean crop, Cambridgeshire. (photo: N M Saville)

(fig. 3). These areas should be protected from ploughing and broad spectrum herbicides, which destroy the perennials, and from fertilizers, which lead to the dominance of rank weeds such as stinging nettle that offer no reward to bees. Unsprayed headlands, as recommended by the Game Conservancy[31], may not in themselves produce bee forage, but they can help to protect established vegetation in field margins from contamination by insecticides, herbicides and fertilizers applied to the crop.

Hedgerows, waysides and woodlands

The arable landscape is a mosaic of cultivated fields in a network of uncropped areas, such as hedgerows, banks, waysides and small farm woodlands. These support perennial vegetation and act as refuges or reservoirs for wild bees and other wildlife[2]. The scale of the mosaic matters. Bumble bees seldom fly more than 100–200 m from the nest to forage[29], and the high costs of long-distance travel must be set against the profit from foraging, so forage is most valuable if it is close to the nest. Bees nesting in a copse in the corner of a field with a continuous network of flowery hedgerows have access to more foraging sites than bees nesting in an isolated wood in the middle of a 20-ha field.

Although bumble bees do not forage in the dense shade of the woodland canopy, sensitively managed farm woodlands can be valuable to bees. In open rides, wood edges, clearings and young plantations, vegetation can be allowed to develop over many years without disturbance of the soil surface, and can therefore provide nesting places for wild bees and forage for bumble bees and honey bees. Bees usually forage in sunny places, not in the shade[6, 29], and occasional mowing may be necessary to maintain open, sunny foraging sites. The cutting regime may have important effects on habitat quality. Summer mowing deprives bees of flowers in a season when their demands are high; mowing in autumn or spring is preferable. Similarly, bees suffer if large areas are cut at one time; a mosaic mowing regime is less likely to interrupt foraging.

Land taken out of arable cultivation

Land taken out of arable cultivation, such as set-aside or the inter-row areas of new plantations, can provide forage for bees if it supports a seasonal succession of nectar-rich flowers (fig. 4). This can be achieved by natural regeneration, by sowing wild flower mixtures, or by sowing bee-forage crops. The wild flower option costs more but may produce a sward of higher diversity in a shorter time than natural regeneration. The choice will depend in part on local circumstances.

FIG. 4. Natural regeneration in newly planted farm woodland. (photo: N M Saville)

Natural regeneration If vegetation is allowed to establish on previously arable land by natural regeneration, forage valuable to bumble bees will take several years to develop. The flowers present in the first year will be predominantly small-flowered annuals. It takes time for biennials and perennials to invade, establish and come into flower. The weed problem is expected to be substantial in year one, and may tempt farmers to destroy the vegetation by ploughing or herbicide treatment. If they yield to this temptation, annual weeds will again present a problem in the following year. If they can resist that temptation, perhaps mowing several times to prevent weed seed set but leaving the vegetation at the soil surface intact, the situation will improve. The annual weed problem will diminish progressively in the second and subsequent years as an increasing diversity of perennials displace the annual weeds[27, 30]. When established, the perennial vegetation should require little management except infrequent mowing to prevent scrub invasion, and perhaps spot treatment of persistent perennial weeds with selective herbicides. As well as forage and nesting places for bees, perennial vegetation provides a home for the natural enemies of pests such as aphids, together with other insects, plants and birds of conservation value.

Sowing wild flower seed Natural regeneration is to be preferred in most cases, but sometimes it is not a viable option. For instance some fertile areas are dominated by rank perennial weeds of productive land. Although some of these, such as rosebay willow-herb, *Epilobium angustifolium*, provide good bee forage, in general they are of low conservation value, and some, such as stinging nettles, not only provide no bee forage at all but also prevent colonization by more valued perennial species. Other sites lack nearby vegetation that can act as a source of seed for natural recolonization by a diverse local flora. In cases like these it may be appropriate to bypass the early stages of natural regeneration by sowing a carefully-designed wild flower seed mixture, consisting of species chosen because they are known to thrive in the region and including broadleaved perennials that are good sources of nectar or pollen for flower-visiting insects. Some suitable species are listed by Fussell and Corbet[15].

It is important that the seeds come from a reliable source that can guarantee their British origin (see table 2 for some suggestions). If wild flower seeds that differ genetically from local forms are sown nearby, existing areas of good natural vegetation may be contaminated by gene flow in the form of seeds or pollen. It is therefore undesirable, as well as unnecessary, to sow seeds if good natural vegetation is available close by. Wild flowers can establish well from seed if the soil is well prepared before sowing. Frequent mowing may be necessary in the first year or two to suppress annual weeds[30], but the perennial sward that eventually develops should soon require less management, and may even offer better bee forage than one established by natural regeneration.

Bee-forage crops If there is no alternative to annual ploughing, sowing an annual crop on set-aside land or elsewhere may be the only way to provide forage for bees. Annuals are important on the farm, both as weeds and as crops, because many agricultural procedures involve annual cultivation. Although most annual arable weeds are small plants with small flowers, and cannot manufacture enough sugar to produce much nectar, there are a few exceptional annuals with large, nectar-rich flowers much visited by social bees. Some of them, such as yellow rattle (*Rhinanthus* spp.)[17] and red bartsia (*Odontites vernus*)[29], are hemiparasites, tapping the roots of a perennial host plant as a source of assimilate, some of which may be sugars destined for the nectar[32]. Others are competitive species with large seeds and rapid early growth that enables them to achieve a large photosynthetic area quickly. This means that they can manufacture abundant nectar sugar for bees, and it also has other advantages on the farm — some large-seeded annuals are already sown to suppress arable weeds and for game cover.

A very few nectar-rich annuals are native in Britain (examples are common hemp nettle, *Galeopsis tetrahit*, and red dead nettle, *Lamium purpureum*), but most originate in other countries. Himalyan balsam, *Impatiens glandulifera*, is an invasive weed, but among these special introduced annuals are a few with promise as bee forage crops. They include borage, phacelia, sunflower, kale, or even perhaps single larkspur. When a patch of phacelia (*Phacelia tanacetifolia*) is in flower, local bees have abundant forage. But the bounty is transient, and when flowering ends the bees may not find enough forage to finish rearing the brood formed during the period of plenty. Bumble bee colonies will suffer if the seasonal succession of forage is interrupted, even for only a few days. The seasonal spread of bee forage can be extended by successional sowings of crops such as phacelia[35], or by cutting patches to stimulate a late second flowering, but ideally even more variety can be introduced by growing a carefully-planned selection of different bee-forage crops, to provide forage in all seasons for all species of bumble bee.

If the agronomic context can accommodate perennials, more options are available. Plant families particularly valuable as bee forage include the Leguminosae and the Labiatae[15]. Legumes such as red clover, white clover, birdsfoot trefoil, sainfoin and melilot are of known value[21, 29]; but set-aside regulations will limit

---

TABLE 2. Some companies that offer wildflower seed of British origin.

Emorsgate Seeds,
Emorsgate, Terrington St Clement, Kings Lynn,
Norfolk PE34 4NY

W W Johnson & Son Ltd,
Boston, Lincs PE21 8AD

---

the proportions in which legumes may be incorporated into seed mixes. The family Labiatae includes species of high value for bees[21]. Current set-aside regulations specifically exclude lavender and sage[18], but other labiates are expected to provide valuable forage that could help to sustain long-tongued bumble bees. Many labiates suitable for bees are crops grown for volatile oils, and are therefore likely to be permitted crops on set-aside land. Their agronomy, processing and marketing are described by Hay and Waterman[16]. Among the labiates they list are a few annuals (summer savory, *Satureja hortensis*, and sweet basil, *Ocimum basilicum*), and numerous perennials including lemon balm, *Melissa officinalis*; hyssop, *Hyssopus officinalis*; rosemary, *Rosmarinus officinalis*; and winter savory, *Satureja montana*. Bergamot, *Monarda fistulosa*, a herbaceous perennial crop grown as a source of geraniol[28], has flowers attractive to long-tongued bumble bees.

Monitoring bumble bee communities

How can a farmer assess the pollinators available on a particular farm, and monitor the success of a programme designed to augment their populations? It is rarely practicable to evaluate pollination directly. Carefully planned experiments are required to discover whether or not inadequate pollination is limiting yield; other agronomic problems may be responsible for a shortfall[5]. It is easier to monitor the bees themselves, for instance by making regular counts, say once a week through the season, of bumble bees seen along a 'bee walk', say 100–500 m long, in a flowering crop or along a field boundary.

One way to do this is to use a method we adopted to explore the role of seminatural vegetation in Cambridgeshire farms[29]. A standard route was broken into sections for recording purposes. Records were made only on fine days, because bumble bees are less active in dull, cool weather. Walks took place at a standard time of day: bumble bees generally forage early and late in the day but are sometimes less active around noon, so walks could be made, say, between 09.00 and 11.00 h. In each section of the walk, the observer noted the species of every bumble bee seen foraging

within 1 m either side of her, and recorded what species of plant it was visiting. The method would need to be adapted for use on particular crops. For instance, field bean flowers close every night and reopen gradually through the day, so the crop offers little bee forage in the mornings. By late afternoon most flowers are open and bee numbers peak[12]. On this crop a bee walk done between 16.00 and 18.00 h BST is more informative than one done in the morning.

Naming wild flowers is not difficult, and there are many good illustrated field guides to help. Naming the bees might seem to present more of a problem, because there are several groups of look-alike bumble bee species that share common colour patterns. This makes it hard to recognize species of bumble bees.

The difficulty is less serious than it might seem. The pattern of recent local extinctions is such that in the intensely arable regions of Britain only one or two species in each colour group remain. This means that we can use a simple naming system based on colour pattern to make the bee counts we need to evaluate the level of bee pollination expected on a particular farm[11, 15, 24, 34]. The colour groups can be identified from the cards, with illustrations and a simple key to colour groups, that were originally used for the public survey of forage plants[1]. Table 3 lists the bumble bee colour groups and table 4 lists some of the wild and cultivated flowers that each uses in each season.

Conclusions and recommendations

Several bumble bee species have already been lost from arable areas of Britain, leaving some crops and wild flowers with only a single species that matches the flowers well enough to act as a pollinator. This means that any further losses of species are expected to create critical gaps in the background pollination service, leaving some wild plants and crops without pollinators.

Regional losses of bee species are probably irreversible, because of a spiral interplay between forage plants and pollinators[3, 20, 26]. If a bee goes, the crops and forage plant species that match it may fail to set seed.

The crops may then be abandoned by farmers, unless they can be grown economically from imported seed. The unpollinated wild flower species may disappear from the landscape. Once its matching forage plants have gone, the bee is unlikely to succeed in re-establishing. If the habitat is to be improved for bumble bees (and honey bees) by augmenting forage through the year to reduce the risk of further extinctions, the following points deserve consideration.

● Bumble bees are valuable pollinators of crops and wild plants. In general, a bumble bee species suitable for a particular crop will require a seasonal succession of other flowers resembling the crop's flowers in some respects (for example in corolla depth) to sustain its colonies through the period when the crop is not in bloom.

● Bumble bees are at risk: several species have already been lost from arable regions of Britain and further losses would create gaps in the pollination service.

● Forage flowers (and nesting places) are found in established stands of open perennial vegetation such as hedgerows, waysides and farm woodland edges and rides. Perennial herbaceous vegetation should be safeguarded, and its destruction by ploughing or herbicide treatment should be avoided.

● Existing areas of such vegetation can be supplemented on set-aside land or between the rows of trees in newly-planted farm woodlands by allowing natural regeneration, or by sowing wild flower seed mixtures or a seasonal succession of selected bee-forage crops.

● When a stand of perennial vegetation is established by natural regeneration, control of annual weeds may require frequent mowing in the first year or two, but in later years perennials will suppress annual weeds. Occasional mowing may then be required to prevent scrub encroachment.

● Mowing in summer is to be avoided, because it destroys forage flowers, and large areas of forage should not be cut all at the same time.

● By monitoring bumble bees in terms of colour groups, a farmer can assess the local abundance of species suitable for particular crops, and track year-by-year changes in the bumble bee community to evaluate the impact of habitat management.

Colour group	Main species	Rarer species
Browns	*Bombus pascuorum*	*B. muscorum, B. humilis, B. distinguendus*, males of *B. subterraneus* and *Psithyrus campestris*
Black-bodied red tails	*B. lapidarius*	*B. ruderarius, P. rupestris, P. campestris*
Banded red tails	*B. pratorum*	*B. monticola, B. sylvarum*, males of *B. lapidarius, B. ruderarius* and *P. rupestris*
Two-banded white tails	*B. terrestris* and *B. lucorum*	*B. soroeensis, B. magnus, B. subterraneus, P. bohemicus, P. vestalis, P. sylvestris, P. campestris, P. barbutellus*
Three-banded white tails	*B. hortorum*	*B. ruderatus, B. jonellus, B. subterraneus, P. sylvestris, P. bohemicus*, male *B. lucorum, P. vestalis, P. barbutellus* and *P. campestris*

TABLE 3. Bumble bee colour groups in Britain, with the common species and some rarer species included in each[15].

TABLE 4. Some flowers that received numerous visits by bumble bees of the five colour groups in the national survey. Some potentially important plants are omitted because they were not found often enough in the survey. Some plants for which particular bees are important do not appear here because the plant may not form a large part of the bee's diet, although the bee may be an important pollinator for the plant. Plant taxa that received at least 20 visits over 10 or more walks are included. A + indicates that the plant received more than 5% of that bee's visits during at least one half month. The three columns for each colour group represent seasons: left, April–May; middle, June–July; right, August–September (see Fussell and Corbet[15] for further details).

Flower	Two-banded white tail			Black-bodied red tail			Banded red tail			Brown			Three-banded white tail		
Dandelion	+	-	-	+	-	-	-	-	-	+	-	-	-	-	-
Red deadnettle	-	-	-	-	-	-	-	-	-	+	-	-	-	-	-
White deadnettle	+	-	-	+	-	-	+	-	-	+	+	-	+	-	-
Chives	-	-	-	+	-	-	+	-	-	-	-	-	-	-	-
Sage	-	-	-	-	-	-	-	-	-	+	-	-	-	-	-
Red clover	-	-	-	-	-	-	-	-	-	-	-	-	+	-	-
Cotoneaster	+	+	-	+	+	-	+	+	-	-	-	-	-	+	-
Rhododendron	+	+	-	+	-	-	-	+	-	-	-	-	+	+	-
Vetches	-	-	+	-	-	-	-	-	-	-	+	+	-	-	+
Raspberry	-	-	-	-	-	-	+	+	-	+	-	-	-	-	-
Comfrey	-	+	-	-	-	-	-	+	-	-	-	-	-	-	-
Foxglove	-	+	-	-	-	+	-	-	-	-	-	-	-	+	+
White clover	-	-	-	-	+	+	-	-	-	-	-	-	-	-	-
Woundworts	-	+	-	-	-	-	-	-	-	-	+	-	-	+	-
Willowherbs	-	+	-	-	-	-	-	-	-	-	-	-	-	-	-
Honeysuckle	-	-	-	-	-	-	-	-	-	-	-	-	-	+	-
Larkspur	-	-	-	-	-	-	-	-	-	-	-	-	-	+	-
Cranesbills	-	-	-	-	+	-	-	-	-	-	-	-	-	-	-
Bramble	-	+	+	-	-	-	-	+	-	-	+	-	-	-	-
Lavender	-	-	+	-	+	+	-	+	+	-	+	-	-	-	+
Campanula	-	-	-	-	+	-	-	-	-	-	-	-	-	-	-
Thistles	-	-	+	-	+	+	-	+	+	-	-	+	-	-	+
Knapweed	-	-	+	-	+	+	-	-	+	-	-	+	-	-	-
Runner bean	-	-	+	-	-	-	-	-	-	-	-	-	-	-	-
Heathers	-	-	+	-	-	-	-	-	-	-	-	-	-	-	+
Dahlia	-	-	-	-	-	+	-	-	-	-	-	-	-	-	-
Sedums	-	-	+	-	-	+	-	-	+	-	-	-	-	-	-
Fuchsia	-	-	+	-	-	-	-	-	-	-	-	+	-	-	-
Michaelmas daisy	-	-	+	-	-	-	-	-	-	-	-	+	-	-	-
Ragwort	-	-	-	-	-	+	-	-	+	-	-	-	-	-	-
Marigold	-	-	-	-	-	-	-	-	+	-	-	-	-	-	-
Himalayan balsam	-	-	-	-	-	-	-	-	-	-	-	+	-	-	-
Buddleia	-	-	-	-	-	-	-	-	-	-	-	-	-	-	+

We conclude that the most important factor in management to conserve bumble bee populations is to safeguard and extend areas of perennial herbaceous vegetation. Annual cultivation destroys the habitats that support bumble bees and many other valued species of animals and plants on the farm.

References

The numbers given at the end of references denote entries in *Apicultural Abstracts*.

1. ANON (1992) *Bumblebee identification card*. Richmond Publishing; Slough, UK.
2. BANASZAK, J (1992) Strategy for conservation of wild bees in an agricultural landscape. *Agriculture, Ecosystems and Environment* 40: 179–192.
3. CORBET, S A (1987) More bees make better crops. *New Scientist* 115: 40–43.
4. CORBET, S A; CHAPMAN, H; SAVILLE, N (1988) Vibratory pollen collection and flower form: bumbe-bees on *Actinidia*, *Symphytum*, *Borago* and *Polygonatum*. *Functional Ecology* 2: 147–155.
5. CORBET, S A; WILLIAMS, I H; OSBORNE, J L (1991) Bees and the pollination of crops and wild flowers in the European Community. *Bee World* 72(2): 47–59.   1449/91
6. CORBET, S A; FUSSELL, M; AKE, R; FRASER, A; GUNSON, C; SAVAGE, A; SMITH, K (1993) Temperature and the pollinating activity of social bees. *Ecological Entomology* 18: 17–30.
7. DELBRASSINNE, S; RASMONT, P (1988) Contribution à l'étude de la pollinisation du colza, *Brassica napus* L. var. *oleifera* (Moench) Delile, en Belgique. *Bulletin des Recherches Agronomiques de Gembloux* 23(2): 123–152.   335/90
8. FEBER, R (1993) *The ecology and conservation of butterflies on lowland arable farmland*. DPhil thesis; Oxford University, UK.
9. FUSSELL, M (1992) Diurnal patterns of bee activity, flowering, and nectar reward per flower in tetraploid red clover. *New Zealand Journal of Agricultural Research* 35: 151–156.
10. FUSSELL, M; CORBET, S A (1991a) Forage for bumble bees and honey bees in farmland: a case study. *Journal of Apicultural Research* 30(2): 87–97.   915/92
11. FUSSELL, M; CORBET, S A (1991b) Bumblebee habitat requirements: a public survey. *Acta Horticulturae* 288: 159–163.   1166/91
12. FUSSELL, M; OSBORNE, J L; CORBET, S A (1991) Seasonal and diurnal patterns of insect visitors to winter sown field bean flowers in Cambridge. *Aspects of Applied Biology* 27:95–99.
13. FUSSELL, M; CORBET, S A (1992) Observations on the patrolling behaviour of male bumblebees (Hym.). *Entomologist's Monthly Magazine* 128: 229–235.
14. FUSSELL, M; CORBET, S A (1992) The nesting places of some British bumble bees. *Journal of Apicultural Research* 31(1): 32–41.   401/93
15. FUSSELL, M; CORBET, S A (1992) Flower usage by bumble-bees: a basis for forage plant management. *Journal of Applied Ecology* 29: 451–465.

16. HAY, R K M; WATERMAN, P G (eds) (1993) *Volatile oil crops*. Longman; Harlow, UK; 185 pp.
17. KWAK, M M (1979) Effects of bumblebee visits on the seed set of *Pedicularis*, *Rhinanthus* and *Melampyrum* (Scrophulariaceae) in the Netherlands. *Acta Botanica Neerlandica* 28(2/3): 177–195.   784/91
18. MINISTRY OF AGRICULTURE, FISHERIES AND FOOD (1993) *Arable area payments 1993/2: explanatory guide part I*. Ministry of Agriculture, Fisheries and Food; London, UK.
19. OSBORNE, J L. Unpublished data.
20. OSBORNE, J L; WILLIAMS, I H; CORBET, S A (1991) Bees, pollination and habitat change in the European Community. *Bee World* 72(3): 99–116.   341/92
21. PATTEN, K D; SHANKS, C H; MAYER, D F (1993) Evaluation of herbaceous plants for attractiveness to bumble bees for use near cranberry farms. *Journal of Apicultural Research* 32(2): 73–79.   229/94
22. PESSON, P; LOUVEAUX, J (eds) (1984) *Pollinisation et productions végetales*. Institut National de la Recherche Agronomique; Paris, France; 663pp.
23. PETERS, G (1972) Ursachen für den Rückgang der seltenen heimischen Hummelarten (Hym., *Bombus* et *Psithyrus*). *Entomologische Berichte* 1972: 85–90.   55L/76
24. PRYS-JONES, O E; CORBET, S A (1991) *Bumblebees*. Richmond Publishing; Slough, UK; 92 pp.
25. RASMONT, P (1988) *Monographie écologique et zoogéographique des Bourdons de France et de Belgique (Hymenoptera, Apidae, Bombinae)*. PhD thesis; Faculté des Sciences agronomiques de l'Etat; Gembloux, Belgique; 370 pp.   1097/89
26. RATHCKE, B J; JULES, E S (1993) Habitat fragmentation and plant-pollinator interactions. *Current Science* 65: 273–277.
27. ROEBUCK, J F (1987) Agricultural problems of weeds on the crop headland. In Way, J M; Greig-Smith, P W (eds) *Field margins*. British Crop Protection Council; Thornton Heath, UK; Monograph No. 35; pp 11–22.
28. ROUGEMONT, G M DE (1989) *A field guide to the crops of Britain and Europe*. Collins; London, UK; 367 pp.
29. SAVILLE, N M (1993) *Bumblebee ecology in woodlands and arable farmland*. PhD thesis; University of Cambridge, UK; 331 pp.
30. SMITH, H; MACDONALD, D W (1992) The impacts of mowing and sowing on weed populations and species richness of field margin set-aside. In Clarke, J (ed) *Set-aside*. British Crop Protection Council; Farnham, UK; Monograph No. 50; pp 117–122.
31. SOTHERTON, N W (1991) Conservation headlands: a practical combination of intensive cereal farming and conservation. In Firbank, L G; Carter, N; Darbyshire, J F; Potts, G R (eds) *The ecology of temperate cereal fields*. Blackwell Scientific Publications; Oxford, UK; pp 373–397.
32. STEWART, G R; PRESS, M C (1990) The physiology and biochemistry of parasitic angiosperms. *Annual Review of Plant Physiology and Plant Molecular Biology* 41: 127–151.
33. TUTTON, M W L (1930) *Diary of natural history for 1930*. Archived in Cambridge University Library, UK.
34. WATCH (1987) *Really useful insects project pack*. Richmond Publishing; Slough, UK.

35. WILLIAMS, I H; CHRISTIAN, D G (1991) Observations on *Phacelia tanacetifolia* Bentham (Hydrophyllaceae) as a food plant for honey bees and bumble bees. *Journal of Apicultural Research* 30(1): 3–12. 523/92

36. WILLIAMS, P H (1986) Environmental change and the distributions of British bumble bees *(Bombus* Latr.). *Bee World* 67(2): 50–61. 39/87

Who cares for solitary bees?

Christopher O'Toole

Hope Entomological Collections, University Museum, Oxford, OX1 3PW, UK

Introduction: the bees

In the context of this paper, the term 'solitary' or 'wild' bees refers to species other than *Apis*. Wild bees are important for two reasons:

- Through their often highly specialized relationships with native floras, solitary bees play vital roles in maintaining natural vegetation in both temperate and tropical regions[19, 21, 26].

FIG. 1. Females of the ground-nesting species *Colletes succinctus* at a nest aggregation in the North Yorkshire Moors. Many solitary bees nest in aggregations, but each nest is the work of a single female.

- The pollination of several crops around the world relies primarily on solitary bees, and an increasing number of wild bee species are actively managed for the purpose, especially in the USA and Japan[20].

Given the dual importance of solitary bees, there is growing concern that modern intensive agriculture and other land use policies result in habitat changes which are hostile to solitary bees.

Solitary and primitively social bees comprise a diverse group in seven families. In true solitary species, each nest is the work of a single female working alone: there is no caste of co-operating sterile females (workers). In the western Palaearctic, primitively social species are found mainly in the Halictidae, in genera such as *Halictus* and *Lasioglossum*. On the basis of nesting behaviour, solitary and primitively social bees can be divided into two broad categories: ground-nesters and cavity- or aerial-nesters. All species provision their cells with a mixture of pollen and nectar and some species also collect plant oils.

Ground-nesters

Sometimes called mining bees, species in this category excavate nest burrows in soil (fig. 1), and sometimes in the soft mortar of old walls. Light, well-drained soils are favoured but many species also prefer friable clay or compacted soil and some can excavate soft chalk and sandstone. Offspring are reared in natal cells lined with a glandular secretion of the female. Genera include *Colletes*, *Andrena*, *Panurgus*, *Halictus*, *Lasioglossum*, *Melitta*, *Dasypoda*, *Macropis*, *Eucera* and *Anthophora*.

Cavity-nesters

These bees nest in ready-made cavities such as hollow plant stems and beetle borings in dead wood. Such species are opportunists and readily nest in artificial structures such as keyholes, old nail holes and crevices in old mortar. For this reason, several species have been both accidentally and deliberately transported around the world by humans. With the exception of species of *Hylaeus*, cavity-nesters do not line their brood cells with glandular secretions: instead,

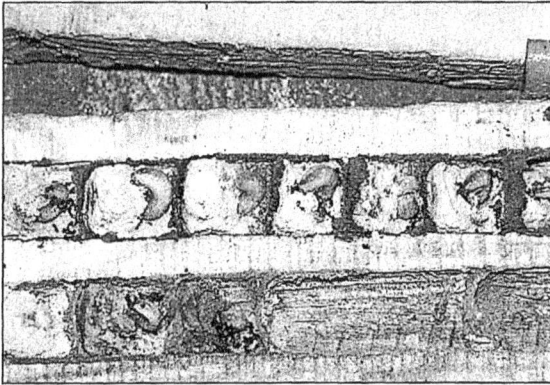

FIG. 2. Two nests of the red mason bee, *Osmia rufa*, in drilled wood, split to show larvae, stored pollen and mud partitions. (photo: P O'Toole)

according to species, they collect and modify materials such as mud (fig. 2), dung, resin, plant and animal hairs, leaf pieces and a mastic of chewed leaves and/or petals. Genera include *Hylaeus*, *Anthidium*, *Heriades*, *Chelostoma*, *Osmia*, *Megachile* and *Ceratina*.

Both categories of bees include cleptoparasitic or 'cuckoo' bees, which lay their eggs in the nests of other species. General accounts of the biology of solitary bees are available[21, 26, 33, 34]. Although bumble bees (*Bombus* spp.) are not part of the remit of this paper, they receive mention where relevant.

Wild bees in decline: the problems?

There is evidence that in Britain and other countries of the European Union populations of wild bees are undergoing decline[10, 20, 35, 36]. An outline for the reasons for this is given in table 1. There is a consensus that modern agribusiness, with its emphasis on capital return, inevitably exerts negative pressure on populations of wild bees. This is via 'improvement' or reclamation of hitherto 'non-productive' land, the elimination of marginal areas in the corners and sides of fields, loss of hedgerows, verges and streamside vegetation. Even where hedgerows are preserved, they are often 'tidied up' so that many hedgerows have the appearance of being given a 'short-back-and sides' haircut.

In a recent government-sponsored report, Barr[7] links the decline of bird, butterfly and plant diversity with increased intensification of agriculture. The report also highlights the puzzling fact that while the rate of loss of hay meadows has declined or even halted in recent years, there is still an on-going loss of floral

TABLE 1. Reasons for declines in populations and extinctions in wild bee species.	
Cause	Effect
Overgrazing by sheep, horses	Reduction of floral diversity, destruction of nest sites
Undergrazing by rabbits	Reduction of floral diversity by proliferation of coarse grasses, loss of nest sites associated with rabbit burrows
Intensive agriculture	Herbicides leading to reduced floral diversity, insecticides resulting in reduced bee populations, loss of marginal lands, loss of major habitats (e.g. Ruddheath, Cheshire)
Intensive agriculture and civil engineering projects	Over-fragmentation of populations (e.g. Twyford Down, Hampshire)
Intrinsic factors	Extreme oligolecty (e.g. *Melitta dimidiata* on *Onobrychis viciifolia*), specialized nest sites (e.g. *Andrena trimmerana* and rabbit burrow entrances)

TABLE 2. Land cover in mainland Britain (percentages of total area calculated from computer analyses of satellite imaging)[a].

Coniferous woodland	3.2
Broadleaved woodland	5.1
Urban areas	6.6
Micellaneous	11.9
Arable land	21.4
Moorland, heath, bracken	24.5
Managed grassland (grazing)	27.3

[a]Source: *The countryside survey 1990.* (Institute of Terrestrial Ecology & Institute of Freshwater Ecology, 1993)

diversity. The report also cites other changes in the six years between 1984 and 1990 which are likely to be harmful to bees: 1% of broadleaved woodland was lost, coniferous plantations increased in area by 5%, the area of land occupied by buildings and roads increased by 4%, and 7% of mainland Britain is now covered by bracken. Table 2 shows the extent to which land use in Britain is largely hostile to solitary bees.

In addition to structural changes, the application of herbicides and the destruction of traditional hay meadows have resulted in a reduction of floral diversity. The same effects are seen in country lanes, where financial pressures have reduced the frequency with which local authorities mow roadside verges: coarse grasses choke out wild flowers.

In summary, adverse structural change impacts on bees via the reduction of nest sites, and reduced floral diversity means less food.

The widespread loss of rabbit populations is a paradigm of both structural change and reduced floral diversity. Many ground-nesting solitary bees nest in the exposed, bare earth surrounding rabbit warrens: some species, such as the now rare *Andrena trimmerana*, specialize in nesting in the vicinity of rabbit burrows. Rabbit grazing creates and maintains flower-rich swards and keeps down coarse grasses.

In discussing and summarizing the constellation of pressures causing declines in British bumble bees,

Williams[35] stresses the difficulty of assessing the relative importance of individual factors. He also suggests that the cooling of the British climate which took place between the 1950s and 1980s may have had adverse effects. His discussions are no doubt relevant to solitary bees.

Solitary bees are able to survive in small, isolated pockets of appropriate habitat, sometimes at low population densities. Else[12, 13] has found several nationally rare solitary bees in land formerly owned and managed by the Ministry of Defence. Among them was *Andrena hattorfiana*, a British Red Data Book species classified as endangered, and recently discovered surviving in a small, isolated population at Dry Sandford Pit, Oxfordshire[18]. This ability is shared with solitary hunting wasps, Sphecidae[11]. While this may be grounds for optimism, it is not without risks: over-fragmented nesting populations are vulnerable to even small, local structural changes. For example, in order to increase grazing pasture for dairy cattle, the destruction of Ruddheath, an area of sandy heath in central Cheshire, took only one week, but involved the loss of several heathland species, particularly a nesting population of *Colletes succinctus*[18].

C. succinctus is oligolectic on species of *Erica* and *Calluna* (Ericaceae) and was vulnerable for that reason: oligolectic bees are specialists on a restricted range of food plants as sources of pollen. Broadly speaking, oligolectic bees may be restricted to particular plants at the level of the family or genus. Thus *Colletes daviesanus* forages only on members of the daisy family, Asteraceae. Narrowly oligolectic species are restricted to either one species of plant or a group of related species, for example *Hylaeus signatus* on species of *Reseda* (Resedaceae), or *Macropis euopaea* restricted entirely to one species, yellow loosestrife, *Lysimachia vulgaris*. While oligolectic bees are often the most efficient harvesters of pollen from their host plants, often by virtue of co-evolved structural modifications, they are vulnerable if their host plant populations are threatened by habitat change or disease. Thus Else[13] reports the disappearance in one season of a formerly thriving Hampshire population of *Hylaeus signatus* resulting from the inexplicable failure to flower of *Reseda lutea* and *R. luteola*.

Twenty-eight species of wild bee are listed as either vulnerable, endangered or extinct in Britain[15] (table 3). This accounts for 11.1% of the British bee fauna.

Oxfordshire

A more general picture of decline in our fauna of native wild bees can be seen by considering the bees recorded for Oxfordshire. This county is well-recorded, with 149 species listed by Richards[25]. Many of the records were made by former members of the Hope Department of Entomology at Oxford University (now the Hope Entomological Collections of the University Museum, Oxford). For the most part, the specimens on which these records are based are in the Hope Entomological Collections and it has been possible to verify or correct the identifications. I have collected bees in Oxfordshire for 26 years (1967–1993) and it is interesting to compare my findings with those of Richards[25], whose records date from c. 1900 to 1938.

TABLE 3. British 'Red data book' bees[15].			
Species	**Family**	**Nest type**	**Status**
Andrena ferox	Andrenidae	ground	endangered
A. floricola	Andrenidae	ground	endangered
A. gravida	Andrenidae	ground	endangered
A. hattorfiana	Andrenidae	ground	vulnerable
A. lathyri	Andrenidae	ground	endangered
A. lepida	Andrenidae	ground	endangered
A. nana	Andrenidae	ground	endangered
A. polita	Andrenidae	ground	? extinct
A. tridentata	Andrenidae	ground	endangered
A. vaga	Andrenidae	ground	endangered
Halictus eurygnathus	Halictidae	ground	? extinct
H. maculatus	Halictidae	ground	? extinct
Lasioglossum laticeps	Halictidae	ground	vulnerable
Dufourea minuta	Halictidae	ground	endangered
D. vulgaris	Halictidae	ground	endangered
Melitta dimidiata	Melittidae	ground	endangered
Stelis breviuscula	Megachilidae	cuckoo	endangered
Osmia inermis	Megachilidae	cavity	vulnerable
O. uncinata	Megachilidae	cavity	vulnerable
O. xanthomelana	Megachilidae	cavity	endangered
Nomada armata	Anthophoridae	cuckoo	endangered
N. errans	Anthophoridae	cuckoo	endangered
N. guttulata	Anthophoridae	cuckoo	endangered
N. sexfasciata	Anthophoridae	cuckoo	endangered
N. xanthosticta	Anthophoridae	cuckoo	endangered
Eucera tuberculata	Anthophoridae	ground	? extinct
Melecta luctuosa	Anthophoridae	cuckoo	endangered
Bombus cullumanus	Apidae	bumble bee	? extinct

TABLE 4. Numbers of bee species recorded from Oxfordshire and their present status.

Status	Species	
	No.	%
Listed in VCH[25], still extant	39	26.2
Common in VCH, now rare	35	23.5
Widespread in VCH, now rare	17	11.4
Rare in VCH, still rare	31	20.8
Added since VCH	20	13.4
Present status uncertain	2	1.3
Extinct	5	3.4
Total	149	

The results of this comparison are outlined in table 4. While 26% of the species recorded by Richards have been recorded in the last 26 years, 23.4% of those which were originally designated as common and 11.4% of those said to be widespread, are now apparently rare. To put things into perspective, 20 species (13.4%) have been added to the county list of 149 species since 1967, although some of these result from changes in county boundaries. The five species thought now to be extinct in Oxfordshire are all bumble bee (*Bombus*) species.

Table 5 gives a breakdown of the number of bees species recorded from my former garden in Botley, Oxford. This is remarkable for such a small area and compares more than favourably with the results of a much longer and more detailed study of a garden in Leicester[22]. The Leicester and Oxford gardens also have a good representation of the total number of species listed for their respective counties (48.1% and 53.9% respectively). It will be seen from table 5 that the bee faunas of both gardens compare well with those of two local Oxfordshire nature reserves, Dry Sandford Pit and Hitch Copse Pit, and the National Nature Reserve at Newborough Warren-Ynys Llanddwyn, Anglesey, all three reserves being notable for the richness of the wasp and bee faunas.

Thus, while solitary bee faunas have clearly declined in the agricultural landscape, significant remnants survive in domestic gardens. This highlights the fact that domestic gardens, with their contrived floral diversity, mosaics of light and shade, together with structural complexity, collectively constitute an important network of unofficial, bee-friendly nature reserves. This is no exaggeration: the total area of domestic gardens in Britain (485 000 ha) occupies nearly 20 times the area of nature reserves administered by local naturalist trusts (24 600 ha)[22, 23].

TABLE 5. Numbers of species of bees from two counties and several localities, according to nesting type. (The figures in brackets are percentages of the total.)

County or locality	Ground nesters	Cavity nesters	Cuckoo bees	Bumble bees	Total
Leicestershire[2]	50 (48.08)	17 (16.35)	29 (27.88)	8 (7.69)	104
Leicester garden[22]	18 (36.00)	12 (24.00)	12 (24.00)	8 (16.00)	50
Oxfordshire	73 (48.99)	23 (15.44)	39 (26.17)	14 (9.40)	149
Shotover Hill (Oxon)	46 (52.26)	10 (11.35)	24 (27.17)	9 (10.23)	88
Bernwood	26 (40.00)	12 (18.46)	17 (26.15)	10 (15.40)	65
Dry Sandford Pit	30 (51.72)	8 (13.80)	12 (20.68)	8 (13.80)	58
Botley Garden (Oxford)	23 (41.07)	11 (19.64)	14 (25.00)	8 (14.29)	56
Hitch Copse Pit (Oxon)	24 (55.81)	5 (11.63)	7 (16.28)	7 (16.28)	43
Newborough Warren (Anglesey)	11 (32.35)	8 (23.53)	6 (17.62)	9 (26.50)	34

Wild bees in decline: why bother?

About 30% of human food is derived from bee-pollinated plants[16], and two thirds of the land area devoted to insect-pollinated crops in the USA is not supplied with honey bee colonies[17]. Thus, two thirds of American crop pollination depends on the adventive services of managed honey bees, feral honey bees and wild bees.

The North American experience shows that solitary bees are an important resource, a pool of potential pollinators for crops[8, 9, 20, 28]. Principal among them is the alfalfa leafcutter bee, *Megachile rotundata*, a Eurasian species which was accidentally introduced into North America. This is a more efficient pollinator of alfalfa (lucerne), the fourth most valuable crop in North America, than the honey bee.

More recently, *Osmia sanrafaelae* has been shown to have potential for alfalfa pollination in desert areas[24], and *O. ribifloris* is managed for pollination of highbush blueberry[29].

In Britain, the solitary mason bee, *Osmia rufa* (fig. 3), has potential as a managed pollinator in orchards and glasshouses. It is a close relative of *O. cornuta* and *O. lignaria propinqua*. The former is a European species introduced and successfully managed for almond pollination in California and the latter, native to North America, is managed for apple pollination[31]. Like these two species, *O. rufa* nests readily in trap-nests and local populations can be enhanced in this way. *O. cornuta* and *O. lignaria propinqua* can forage at lower temperatures than the honey bee can. My impression that this is also true for *O. rufa* is supported by Stone[27], who showed that the species is endothermic and, for its body mass, has a very high warm-up rate. For a detailed, up-to-date review of the role of solitary bees in crop pollination, see O'Toole[20].

While the honey bee may be the managed pollinator of choice for a variety of crops, it is often inefficient[32]. There are three main reasons for under-pollination by honey bees:

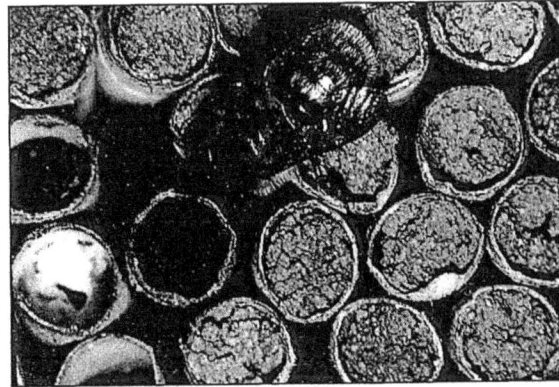

FIG. 3. A female *Osmia rufa* seals her nest with mud, in a trap-nest made of drinking straws which has already been used by several females. (photo: P O'Toole)

FIG. 4. A patchwork of old fields, hedgerows, copses and woods: an older, bee-friendly agricultural landscape in Herefordshire.

- The crop area is too large for the available honey bee colonies, for example, almonds in North America and Israel.

- Dis-incentives for beekeepers arise out of the cost of transporting colonies to offset this factor.

- Honey bees are not attracted to the crop or are physically maladapted to pollinate it. Examples, worth billions of dollars per year in the USA, are alfalfa, soya beans, cotton and sunflower, each of which is adapted to specific types of non-*Apis* pollinators.

The restrictions on moving honey bee colonies, imposed as a result of the pandemic of varroa disease on both sides of the Atlantic, severely limit or even prohibit the pollination services of migratory bee-keeping. For this reason alone, the importance of wild bees as pollinators is thrown into high relief.

Wild bees in decline: what can we do?

Any prescriptions to improve the lot of our solitary, wild bees must centre on attempts to make patterns of land use more bee-friendly (fig. 4)[3,4,5,14]. We can capitalize on the fact that domestic gardens are already in this state: individuals manage their gardens as havens for bees by maintaining a diversity of flowering plants attractive to bees, and they can provide artificial nests (trap-nests) for cavity-nesting species. These nests comprise either bundles of bamboo or drilled blocks of timber, hung in south or south-west facing, sheltered but not shaded situations.

On a larger and national scale, we need to re-establish hedgerows and marginal land as areas of floral diversity and structural complexity (fig. 5). Banaszak and Manole[6] recommend that about 25% of the 'agricultural landscape' should be set aside and managed as flower-rich refuge habitats which would also harbour a range of nest sites. Inevitably, this would require cultural change among farmers, growers and in big agribusiness: they will need persuading that it is not necessary to squeeze the land so hard, that the maximum possible yield is not always desirable. The notorious grain, butter and beef mountains of the EU are ample evidence of this; perhaps farmers could be persuaded that they can afford to invest some of their yield in rabbits, which, if encouraged, will do much of their bee-oriented management work for them.

The recent EU policy of set-aside has potential for enhancing populations of solitary bees: farmers are paid a subsidy for growing cereals if they take 15% of their land out of production, and they also receive payment for each hectare maintained as set-aside land. According to Andrews[1], nearly everywhere wildlife has benefited from set-aside, with increased floral diversity being the most visually immediate

FIG. 5. Marginal land alongside an old drove-way in Gloucestershire: benefiting bees through floral diversity as well as the structural diversity which provides nest sites. (photo: K G Preston-Mafham)

effect as seed banks of arable weeds are allowed to germinate: as many as 243 species of flowering plants appeared in some fields in the first year of set-aside.

However, as Andrews[1] points out, many farmers view this as an eyesore and are, in any case, under considerable peer pressure to maintain 'clean land'. Moreover, many farmers feel negatively towards set-aside. This is understandable in the light of previous incentives and pressures to maximize areas under cultivation and their yields. They will need encouragement to see things differently, to take a pride in the sympathetic management of land for wildlife and increased biodiversity. Indeed, such management would require traditional farming skills of a high order.

In addition to set-aside, the parallel Countryside Premium Scheme (CPS) has potential for encouraging the enhancement of solitary bee populations. Under CPS, farmers receive payments for land specially managed for the benefit of wildlife. Perhaps the rules and regulations governing CPS could be modified to encompass the habitat requirements of wild bees, namely, increasing floral diversity to enhance forage and structural diversity to enhance nest sites.

But cultural change of this sort requires a body of convincing facts, convincingly deployed in a programme of re-education. And this would be easier if we could produce a more solid body of evidence to support the urgency of all these claims. To this end, we need in the short term, sponsored studies on the effects of set-aside on bee populations. In the long term, Europe needs at least one facility on the lines of the USDA Bee Systematics and Biology Laboratory in Logan, Utah. There, USDA scientists study the nesting biology of potential wild bee pollinators, their floral relationships and their experimental use as managed pollinators. They also study the taxonomy or systematics of the bees. In Europe, we have no such centre. And it is urgently needed: we still have no comprehensive taxonomic monographs of western Palaearctic bees, so they remain difficult to identify. The biology of most of them is largely unknown: this means that their conservation needs are sometimes unknown, and their potential as managed crop pollinators is not being realized. We also need more studies of pollination biology at the community level, of the sort being pioneered in Israel[19].

Most countries maintain and update a detailed inventory of their important natural resources: the area of forests, the distribution of minerals, and oil and natural gas reserves. Wild bees, too, are an important natural resource: for Britain, the Bees, Wasps and Ants Recording Scheme (BWARS) attempts to inventory our solitary bees, but we need a more concerted effort than that which BWARS can muster. Indeed, a co-ordinated, pan-European effort, embracing biogeography, biology and systematics is now vital. We neglect wild bee faunas at our peril.

References

The numbers given at the end of references denote entries in *Apicultural Abstracts*.

1. ANDREWS, J (1992) Some practical problems for set-aside management for wildlife. *British Wildlife* 3(6): 329–336.
2. ARCHER, M E (1990) The aculeate solitary wasps and bees (Hymenoptera: Aculeata) of Leicestershire. *Transactions of the Leicester Literary and Philosophical Society* 84: 9–25.
3. BANASZAK, J (1983) Ecology of bees (Apoidea) of agricultural landscape. *Polish Ecological Studies* 9(4): 421–505.
4. BANASZAK, J (1985) Impact of agricultural landscape structure on diversity and density of pollinating insects. *INRA Publications, Paris* 36: 77–84.
5. BANASZAK, J (1990) [An appeal for the legal protection of all species of wild Apoidea.] *Chronmy Przyrode Ojczysta* 1: 5–8 (original in Polish). 144L/91

6. BANASZAK, J; MANOLE, T (1987) [Diversity and population density of pollinating insects (Apoidea) in the agricultural landscape of Romania.] *Polskie Pismo Entomologiczne* 57(4): 747–766 (original in Polish). 358/89
7. BARR, C (1993) *The countryside survey*. Institute of Terrestrial Ecology & Institute of Freshwater Ecology.
8. BOHART, G E (1970) Should beekeepers keep wild bees for pollination? *American Bee Journal* 110(4): 137. 492L/71
9. BOHART, G E (1972) Management of wild bees for the pollination of crops. *Annual Review of Entomology* 17: 287–312. 277/73
10. DAY, M C (1991) *Toward the conservation of aculeate Hymenoptera in Europe*. Council of Europe; Strasbourg, France; Nature and Environment Series No. 51: 44 pp.
11. DOLLFUSS, H (1988) Faunistische Unterschungen uber die Brauchbärkeit von Grabwespen (Hymenoptera, Sphecidae) als Umweltindikatoren durch Vergleich neuer un älterer aufnahmen von ausgewählten Localfaunen in östlichen Österreich. *Linzer Biologische Beitrage* 20(1): 3–36.
12. ELSE, G R (1993) Wildlife reports: ants, bees and wasps. *British Wildlife* 4(6): 295–396.
13. ELSE, G R (1993) Wildlife reports: ants, bees and wasps. *British Wildlife* 5(1): 55–56.
14. ELSE, G R; FELTON, J; STUBBS, A (1978) *The conservation of bees and wasps*. Nature Conservancy Council; London, UK; 16 pp. 830L/79
15. ELSE, G R; SPOONER, G M (1987) Hymenoptera: Aculeata the ants, bees and wasps. In Shirt, D B (ed) *British Red Data Books: 2. Insects*. Peterborough, UK; pp 259–293.
16. MCGREGOR, S E (1976) *Insect pollination of cultivated crop plants*. US Department of Agriculture; Handbook No. 496; 411 pp. 1088/77
17. MCGREGOR, S E; LEVIN, M D (1979) Bee pollination of agricultural crops in the USA. *American Bee Journal* 110(2): 48–50.
18. O'TOOLE, C. Unpublished data.
19. O'TOOLE, C (1991) Wild bees, systematics and the pollination market in Israel. *Antenna* 15(2): 66–72.
20. O'TOOLE, C (1993) Diversity of native bees and agroecosystems. In LaSalle, J; Gauld, I D (eds) *Hymenoptera and biodiversity*. Natural History Museum London & CAB International; Wallingford, UK; pp 169–196.
21. O'TOOLE, C; RAW, A (1991) *Bees of the world*. Blandford; London, UK; 192 pp. 742/92
22. OWEN, J (1991) *The ecology of a garden: the first fifteen years*. Cambridge University Press; Cambridge, UK; 403 pp.
23. OWEN, J; OWEN, D F (1975) Suburban gardens: England's most important nature reserves? *Environmental Conservation* 2: 53–59.
24. PARKER, F D (1985) A candidate legume pollinator, *Osmia sanrafaelae* Parker (Hymenoptera: Megachilidae). *Journal of Apicultural Research* 24(2): 132–136. 342/86
25. RICHARDS, O W (1939) *The Victoria county history of Oxfordshire*. Clarendon Press; Oxford, UK.
26. ROUBIK, D W (1989) *Ecology and natural history of tropical bees*. Cambridge University Press; Cambridge, UK; 514 pp. 1037/90
27. STONE, G N (1990) *Endothermy and thermoregulation in solitary bees*. DPhil thesis; Oxford University, UK.

28. Torchio, P F (1987) Use of non-honey bee species as pollinators of crops. *Proceedings of the Entomological Society of Ontario* 118: 111–124.   693/89

29. Torchio, P F (1990) *Osmia ribifloris*, a native bee species developed as a commercially managed pollinated of highbush blueberry (Hymenoptera: Megachilidae). *Journal of the Kansas Entomological Society* 63(3): 427–436. 1098/91

30. Torchio, P F (1990) Diversification of pollination strategies for U.S. crops. *Environmental Entomology* 19(6): 1649–1656.   698/93

31. Torchio, P F; Asensio, E; Thorp, R W (1987) Introduction of the European bee, *Osmia cornuta*, into Californian almond orchards (Hymenoptera: Megachilidae). *Environmental Entomology* 16: 664–667.   341/89

32. Westerkamp, C (1991) Honeybees are poor pollinators – why? *Plant Systematics and Evolution* 177: 71–75.

33. Westrich, P (1989) *The wild bees of Baden-Württemberg*. Eugen Ulmer; Stuttgart, Germany; pp 5–431.   381/90

34. Westrich, P (1989) *The wild bees of Baden-Württemberg*. Eugen Ulmer; Stuttgart, Germany; pp 437–972.   381/90

35. Williams, P H (1989) *Bumble bees and their decline in Britain*. Central Association of Beekeepers; Ilford, UK; 15 pp. 60L/91

36. Williams, P H (1989) Why are there so many species of bumble bees at Dungeness? *Biological Journal of the Linnean Society* 101: 31–34.   74/91

Use of the Tübingen mix for bee pasture in Germany

Wolf Engels; Ulrike Schulz; Marianne Rädle

Zoologisches Institut, Lehrstuhl Entwicklungsphysiologie, Auf der Morgenstelle 28, D-72076, Tübingen, Germany

Introduction

In Germany modernization of agriculture resulted in a dramatic reduction in bee pasture on arable land. In the past beehives were commonly found on nearly every farm: this is not the case nowadays because between the end of the spring blossom (of cherry, pear and apple trees), and the end of May or beginning of June, only rape blooms and eventually some sunflower. Therefore, honey bee colonies have to be fed in summer at many sites. Pollen sources are particularly lacking. Consequently, today most beekeepers have their apiaries in villages or even big cities, and few colonies are seen in the countryside. Depending on the region and unpredictable weather conditions, some beekeepers practise early autumn migration into the forests for honeydew flows, and into the northern moorlands for heather blossom.

This situation made us look for possible improvements in bee forage on arable land, especially as part of integrated crop management. Research was started by our laboratory nearly 20 years ago, and in the Lautenbach Project we evaluated the significance of: not mowing the green of field paths and ridges; sowing flowers on field margins, inter-crop strips and fallow land; replanting hedgerows and small woods in the field; and sowing undercrops like clover together with cereals or intercrops such as other legumes[1, 2, 3, 10, 14].

In 1987 the European Commission set-aside land programme was started to reduce the overproduction of some food crops. In Germany we began to gain experience with sowing mixtures of annual flowers[4]. There were some preconditions given by the EC, and others were set by ourselves. The fields laid fallow were not allowed to be used for any type of food production, including fodder. Because of the soil nitrate overload, no legumes could be planted. The green cover had to suppress weeds, especially their reproduction, and the sown flowers should not raise problems for the farmer in subsequent cultivation programmes.

We wanted to compose a blooming mix without introducing more new plant species. The main goal was to close the big summer gap in bee pasture, and to provide a long-lasting kaleidoscope of flowers with different types of calyces to be visited by insects of a large range in body size. A good pollen supply, particularly during late summer and autumn, was wanted[5]. Since the EC rules allowed fields to be set aside for either one or five years, and most of the farmers preferred the year-by-year rotation, we decided to use only annual flowers, favouring field herbs, blooming crops and traditional countryside pot-herbs[6, 7]. Some 40 plant species were tested, and the result is the Tübingen mix (Tübinger Mischung)[9, 14] consisting of the 11 flowers listed in table 1.

	TABLE 1. Composition of Tübingen mix.	
Proportion of seed mix (% by weight)	Plant species[1]	Forage provided[2]
40	*Phacelia tanacetifolia*, phacelia, Büschelschön	N + P
25	*Fagopyrum esculentum*, buckwheat, Buchweizen	N
7	*Sinapis alba*, mustard, Gelbsenf	N + P
6	*Coriandrum sativum*, coriander, Koriander	P
5	*Calendula officinalis*, ring daisy, Ringelblume	P
5	*Nigella sativa*, black cumin, Schwarzkümmel	P
3	*Raphanus sativus*, oil radish, Ölrettich	N + P
3	*Centaurea cyanus*, cornflower, Kornblume	P
3	*Malva sylvestris*, wild mallow, Wilde Malve	P
2	*Anethum graveoleus*, dill, Dill	P
1	*Borago officinalis*, borage, Borretsch	N

[1]Scientific name, followed by English and German common names [2]Source of nectar (N), pollen (P) or nectar and pollen (N + P)

Development of the Tübingen mix

Most of our field experiments were run in the vicinity of Tübingen on the farm Eckhof at Kreßbach (48° 30′ N, 9° 3′ E) located 463 m above sea level. The fields are surrounded on two sides by forest (fig. 1), consisting mainly of spruce (*Picea excelsor*). Groups of five colonies of honey bees (*Apis mellifera carnica*) in Deutsch Normal magazine hives were placed on stands at the field margins (fig. 2). In 1993 an additional 40 boxes containing nesting materials for wild bees (fig. 3) were distributed in the field and in the edges of adjacent forest. These consisted of six pairs of wooden boards, each containing 12 holes of either 2, 4 or 6 mm diameter.

The development and the honey yield of the honey bee colonies was monitored throughout the season[3], as was the acceptance of the artificial nesting substrates by wild bees, including the number of brood cells provisioned per female. Pollen was collected with entrance traps in the hives, from pollen stores in the combs, from honey sediments, and from samples taken from the brood provisions of the wild bees.

By pollen analysis and quantification[3] the contribution of the different components of the Tübingen mix, and of wild herbs in the field and in the surrounding habitats, to the nutrition of the particular bee species was calculated. The foraging bees and other insects visiting the flowers in the fields were recorded, in part along transects and on square patches[6, 7, 8, 9]. At the same time the development of the vegetation on the fields was recorded.

In 1992–1993 we tried to use Tübingen mix for a two-year-set-aside scheme, and three variations were tried: seed mix sown in 1992, and the ground allowed to regenerate in 1993; seed mix sown in 1992 and 1993 in sequence; seed mix sown in 1993, and in 1993 the field harrowed early in May.

It was interesting to notice that the third treatment resulted in nearly exactly the normal Tübingen mix vegetation. In all these treatments there was no cultivation before May of the second year, and unlimited natural seed production was possible. Over winter many birds fed on these seeds. The remnants of the

FIG.1. Aerial view of the experimental field sown with Tübingen mix, taken on 1 August 1993 from about 150 m above ground level.

vegetation turned out to be used by numerous insects and other invertebrates as hibernating sites. We also have to point out the fact that many wild bee species are ground-nesting, and the overwintering females need undisturbed soil.

FIG. 2. Hives on one of the experimental stands located at the edge of a field with Tübingen mix.

Characteristics of the Tübingen mix

The first component of the Tübingen mix to bloom is buckwheat (fig. 4), producing flowers especially rich in nectar only three weeks after sowing. It is important to mention that there are several buckwheat cultivars, of which only a few are attractive to bees. The next components to bloom are the crucifers, flowering after about one month followed by phacelia blooming after about six weeks: all offering nectar as well as plenty of pollen. Borage flowers appear at the same time as phacelia but bloom for 5–6 weeks longer.

From mid-July until the end of the season wild mallows, cornflowers and ring daisies are seen in permanent blossom, all offering mainly pollen. Coriander, black cumin and dill flowers come later and bloom for only 4–6 weeks. The Tübingen mix provides a continuum of flowers (figs 5 and 6), from the period of rape blossom until the season is ended by autumn frosts.

In our experimental fields, in addition to the 11 sown Tübingen mix flowers, numerous wild herbs also appeared. In 1992, for instance, we counted over 50 species, and in 1993 on the plots with biannual vegetation over 60 herb species. Many of these were well visited by honey bees and wild bees.

FIG. 3. Experimental nesting box for wild bees.

Effect on honey bee colonies

The pollen sources offered by several components of the Tübingen mix throughout the summer result in a very satisfactory development of bee colonies. The summer honey yield was about 5 kg/colony/month, and the colonies never had to be supplied with candy. Nucleus colonies moved in June to the Tübingen mix fields became strong colonies by the end of August. From the pollen traps we know that crucifers were the main pollen loads carried in June and July, followed by cornflower and phacelia. From surrounding habitats the honey bees also collected a lot of white clover pollen. In some years we also added Persian clover (*Trifolium resupinatum*) and sunflower (*Helianthus annuus*) in varying percentages to the Tübingen mix. Both provide good pollen supplies, especially during summer.

Other beneficial effects

On the Tübingen mix fields a great number of bumble bees and other wild bee species were observed. The situation recorded in 1993 on the biannual plots is given as an example. We counted a total of 58 wild bee species of which 35 (table 2) foraged on Tübingen mix flowers and 19 on wild herbs. Only five species were detected in the experimental nesting facilities (fig. 7) but not seen in the vegetation.

In addition to these many wild bee species, numerous other insects visited the fields with Tübingen mix and were seen on flowers there. It is interesting to note that many beneficial insects appeared, such as ladybirds (coccinellids), hover flies (syrphids) and green lacewings (chrysopids). Of the other insect orders a lot

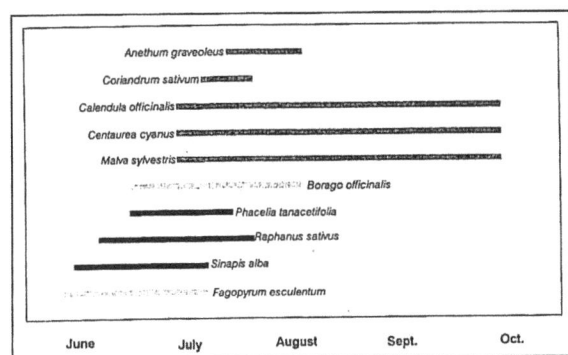

FIG. 4. Flowering sequence of 10 components of the Tübingen mix. Plants are shown as principally sources of nectar (light grey bars), pollen (dark grey bars) or both (black bars). Seed mix sown 7–8 May 1992.

FIG. 6. Detail of Tübingen mix flowers with phacelia, ring daisy, coriander, cornflower and wild mallow, 15 July 1993.

FIG. 5. Experimental field with flowering Tübingen mix, 24 June 1993.

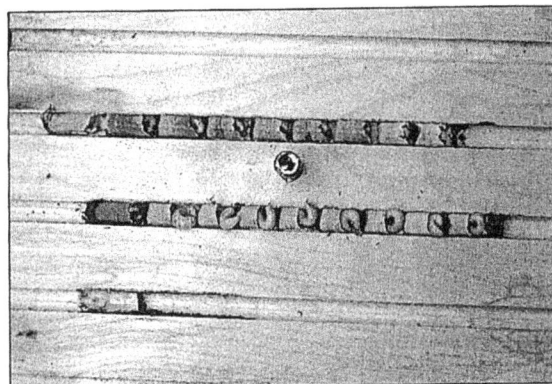

FIG. 7. Nesting box for wild bees, opened to show larvae of *Chelostoma florisomne* (22 June 1993).

TABLE 2. Wild bee species observed foraging, or nesting in artificial substrates, in an experimental field sown with Tübingen mix (1993).

Species	Occurrence	Time of observation
Andrena bicolor	rare	July–August
A. dorsata	very rare	June
A. flavipes	abundant	May, July–August
A. fucata	rare	June
A. minutula	abundant	May, July–August
Anthophora retusa	very rare	July
Bombus distinguendus	very rare	June–July
B. hortorum	rare	June–July
B. humilis	rare	June–September
B. hypnorum	very rare	July
B. lapidarius	abundant	June–September
B. lucorum	rare	June–August
B. pascuorum	abundant	June–September
B. pratorum	rare	June–July
B. soroeensis	abundant	July–August
B. sylvarum	rare	June–August
B. terrestris	rare	June–September
Ceratina cyanea	very rare	August
Chelostoma florisomne	rare, in trap nests only	May–June
Halictus rubicundus	very rare	June–July
Hylaeus communis	abundant	June–September
H. confusus	rare	June–September
H. gredleri	very rare, in trap nests only	June–July
H. kahri	very rare, in trap nests only	July
Lasioglossum calceatum	abundant	July–September
L. leucozonium	very rare	June–August
L. malachurum	abundant	June–September
L. morio	very rare, in trap nests only	September
L. pauxillum	abundant	June–September
L. zonulum	abundant	June–September
Megachile centuncularis	very rare	September
Osmia brevicornis	very rare, in trap nests only	June
Psithyrus barbutellus	very rare	July
P. campestris	very rare	June–July
P. norvegicus	very rare	June
P. rupestris	very rare	July
P. vestalis	very rare	June–July
Sphecodes crassus	very rare	July
Stelis signata	very rare	June–August

of butterflies including several endangered species were eyecatching visitors (fig. 8).

Some other appreciable effects of using Tübingen mix as a blooming green cover for set-aside fields can be mentioned here only briefly: no accumulation of nitrate in the soil, production of much vegetable mould, favourable microclimate, no weed problems, refuge for many invertebrates and small vertebrates.

Seed mixture

The seed mixture for the Tübingen mix has been available on the German market since 1990. The price ranges around DM 7.00/kg (less than 4 ecu/kg), depending on the volume ordered. For medium-quality soil the sowing rate is 7 kg seed mix/ha when using normal sowing machines. It is important to cover the seeds with 2–3 cm of soil because some components, especially phacelia, need to germinate in the dark. The soil should also not be too dry, in order to provide a suitable seed bed. In southern Germany we are sowing the Tübingen mix at the end of the apple blossom, usually 5–10 May.

Implementation

The Tübingen mix was developed to make use of the unique chance of bringing 10–20% of all arable land in Germany, or even the EU, into bee forage. The concept of a bee pasture composed of about 10 annual flower species, of which some will always be in blossom during summer and autumn has now been evaluated and improved over 6 years[14]. Our overall result is that the Tübingen mix blossom is well accepted not only by honey bees but also by many wild bees[8] and other insects, including important beneficial species, thus assisting managed honey bees and the wild insect fauna[13]. The number of wild bee species observed in fields with Tübingen mix makes up approximately 15% of all the bees of Baden-Württemberg[16].

The area sown with Tübingen mix every year now amounts to many thousands of hectares, with not only beekeepers and farmers, but also the public and media such as TV recognizing the blooming fields very

FIG. 8. Swallowtail (*Papilio machon*) feeding on nectar from a phacelia flower.

well. Recreational users of the countryside also enjoy the flowering fields. The beekeepers in the state of Baden-Württemberg actively helped in promoting this new bee pasture, and the state government supported the efforts[15] by sponsoring seed mix costs by DM 200 000/year. The state beekeepers associations ordered the Tübingen mix seeds, and the local beekeepers made arrangements with farmers participating in the set-aside land programme (which until 1991–1992 was a voluntary scheme). The seeds were given to the farmers free of charge, and the beekeepers together with our research group observed the vegetation and the use of forage made by the bees[7]. The Tübingen mix is recommended by the state agriculture department, and the feedback from the farmers is very positive. In fact, sowing of Tübingen mix is the cheapest option for greening set-aside fields, as the seed mix costs only about DM 50/ha. Tübingen mix is listed by the German wholesale seed traders, and over 30 tonnes per year are ordered. Export into neighbouring EU countries is beginning.

Our laboratory has experienced Tübingen mix only in the state of Baden-Württemberg[6,7], which is southwestern Germany. Other field trials in areas with sandy soil, like Brandenburg around Berlin, have shown that not all components of our seed mixture give good and lasting blossom. Evidently this depends on both local climate and soil conditions, and we are very interested to learn how Tübingen mix would work in other European countries. Our intention was to fill the dearth period for bees and other flower-visiting insects which from July until October need a succession of food, especially pollen sources, for reproduction. In other parts of Europe this forage situation may be rather different[12].

The future

Of course even a sown mixture of more than the 11 annual flowers in the Tübingen mix would not provide all the different bee pasture originally present in Europe. But what are the other options, and what benefit can be expected from different measures? The experience with natural green regeneration of set-aside land is negative, at least in Germany. The developing flora may be good bee pasture, but farmers have to cut down the green crop before the weeds reproduce. Mowing once or twice during the summer or even ploughing the fields in June, as recommended in 1993 by the EC, of course destroys all bee forage completely. At a time when many wild bees are actively laying, this would suddenly present the foraging females with empty food dishes. Perhaps a perennial green cover for fields laid fallow for five or even 20 years could also comprise more floral elements of significance for endangered wild bees, including the oligolectic species. The set-aside land policy of the EU covers such solutions, but at least in Germany farmers up to now have hesitated to decide on these long-term contracts. On the other hand, a general recommendation for extensification programmes, with appropriate compensation for the farmers, would have many advantages too[12].

The set-aside land policy of the EU may be a unique chance to make use of a considerable part of the arable fields as new bee pastures[11]. However, is this really a new situation? The old medieval three-field system (Germanische Dreifelderwirtschaft) already included laying fallow the arable land every third year[9]. In Germany the original rotation scheme was winter rye (for bread) first, then summer barley (for beer), followed by a year of natural regeneration. This set-aside land was excellent pasture for cattle and bees, and allowed a constant honey yield on every farm, needed for bread and mead. Strictly speaking we are actually in part returning to the proven worth of our grandfathers' farming schemes.

Our vegetation records on Tübingen mix fields demonstrate how rapidly a multi-species wild herb association can regenerate. Set-aside management should at least provide chances for development and reproduction of these field herbs, which at the same time are food for many oligophagous insect larvae and forage for specialists such as monolectic or oligolectic bees. From the recorded association of the sown flowers, the developing wild herbs and the fauna supported, the biannual rotation scheme of set-aside land is evidently the optimal solution. This is true from the perspective of both the farmer and the ecologist: only about DM 50/ha seed expenses for two years, sowing in the first, and harrowing in the second spring, means minimum costs and labour. An undisturbed piece of land everywhere, with a succession of many flowers including wild herbs, may contribute much to the future conservation of biodiversity in Europe, a continent with agriculture already occupying by far most of the still green environment.

The Tübingen mix is just an attempt to make use of set-aside land for new bee pastures. Our experiences with a biannual scheme of rotational placement of complete fields laid fallow look particularly promising. The many wild herbs in addition to the sown flowers create a remarkably diverse environment. Other seed mixes may be imagined. Other modes of distribution of the set-aside plots within areas of arable land, in particular in form of wide field margins[11], may be even more effective in assisting our indigenous flora and fauna to survive in the European mega-agroecosystems. Additional measures[14] can provide more permanent bee pastures. Bees are indeed of basic significance in the maintenance of biodiversity in general, because their key functions as the main pollinators cannot be replaced.

Any success in the arrangement of new bee pastures within the framework of the EU set-aside policy will depend on initiatives and permanent activities of those interested in bees, both honey bees and wild bees. Therefore, beekeepers in co-operation with farmers have to operate at the local level. Conservationists and bee scientists can and must support all this, and convince decision-makers in politics and administration to finance approved measures. This, of course, has to be done in the future within the continental community of the EU. IBRA will have to play an important role in distributing the relevant information.

Acknowledgements

Our research funds were meritoriously granted by the Ministry of Agriculture and Forestry, State of Baden-Württemberg, Stuttgart. The help of our staff beekeepers, technicians and many student assistants is gratefully appreciated.

References

The numbers given at the end of references denote entries in *Apicultural Abstracts*.

1. BAUER, M (1983) Bienenhaltung in der Feldflur. *Imkerfreund* 38: 412–416.
2. BAUER, M (1985) Verbesserung der Trachtsituation für Bienenvölker in der Feldflur. *Bienenpflege* 1: 7–14. 1222/86
3. BAUER, M (1987) *Bienenweide in der Feldflur: Maßnahmen zur Trachtverbesserung und die Trachtnutzung durch* Carnica-Völker. AS-Verlag; Tübingen, Germany; 292 pp.
4. BAUER, M (1990) Flächenstillegungen – eine Chance zur Verbesserung der Bienenweide? *Bienenpflege* 4: 103–108.
5. BAUER, M (1991) Preliminary results of sowing plants as pasture for bees on former ploughland [summary]. *Apidologie* 22(4): 428–430. 529/92
6. BAUER, M; ENGELS, W (1991) Bienenweide auf stillgelegten Ackerflächen. Tübinger Feldversuche 1990. *Allgemeine Deutsche Imkerzeitung* 25: 40–43; *Bienenpflege* 4: 124–129.
7. BAUER, M; ENGELS, W (1991) Bienenweide auf stillgelegten Ackerflächen. Erfahrungen und Fördermaßnahmen im Land Baden-Württemberg. *Neue Bienenzeitung* 105: 593–595, 656–658.
8. BAUER, M; ENGELS, W (1992) The utilization of the pasture for bees on former ploughland by wild bees [summary]. *Apidologie* 23(4): 340–342. 378/93
9. BAUER, M; ENGELS, W (1992) Bienenhaltung – ein Relikt aus alten Zeiten oder ein Bereich mit neuen Möglichkeiten? Maßnahmen zur Trachtverbesserung im Rahmen des Integrierten Pflanzenschutzes. *Die Neue Bienenzucht* 19: 9–11.
10. GRIESOHN, G (1984) Mehr Brot für die Bienen. *Allgemeine Deutsche Imkerzeitung* 18: 44–48, 50, 79–82, 120–125, 149–151. 1261L/85
11. MATHESON, A (1994) Pastures new. *Bee World* 75(1): 43–46.
12. PAXTON, R (1993) All change down at the farm: a potential for bees and beekeeping. *Bee World* 74(4): 214–220. 223/94
13. RÄDLE, M (1993) Pressekonferenz: Bienenweide auf stillgelegten Ackerflächen. *Bienenpflege* 10: 291–292.
14. RÄDLE, M; ENGELS, W (1993) *Die Bienenweide in der Kulturlandschaft; Landwirte gestalten Lebensräume.* Folder; Reg. Präs; Tübingen, Germany.
15. WEISER, G (1987) Landschaft 'bienenfreundlich' gestalten. *Allgemeine Deutsche Imkerzeitung* 21: 156.
16. WESTRICH, P (1989) *Die Wildbienen Baden-Württembergs.* Ulmer Verlag; Stuttgart, Germany; 972 pp. 381/90

Encouraging bee forage: What can be done in practice?

James Clarke; Kathleen Raw

ADAS, Boxworth Research Centre, Boxworth, Cambridgeshire, CB3 8NN, UK.

Introduction

In the UK in June 1993, there were about 18.5 million hectares of land in agricultural holdings[8], including rough grazing of about 5.8 million ha. Approximately 4.6 million ha was in cropping, 1.6 million ha in temporary grass and 5.2 million ha in permanent grassland. The major arable crop was winter wheat (fig. 1). 1993 was the first year with significant amounts of set-aside land: about 600 000 ha in the UK[10]. The opportunities for bee forage come from three major sources: the existing land use; the opportunities offered by set-aside to manage land specifically for the benefit of bees; and other aspects of general land management.

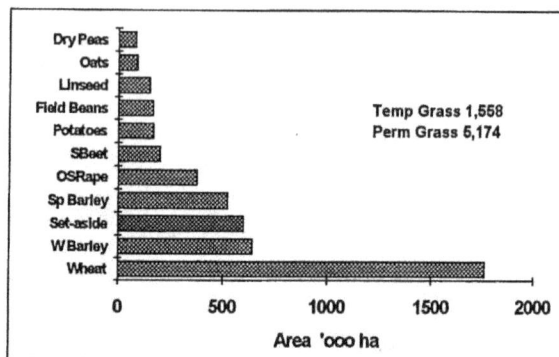

FIG. 1. Cropping in the UK, June 1993 (source: MAFF census).

What can be done for bees
Existing arable area

Existing cropping contains many crops which are favoured by bees. There are already large areas of oilseed rape (fig. 2) and field beans which are likely to remain static.

Existing arable land can be enhanced in its value for bees by the introduction of a wider range of favourable crops, or by more favourable management of existing crops. The area of sunflowers (fig. 3), for instance, has increased in recent years, but still only accounts for a small area (estimated at 600 ha in 1992, and 1 600 ha in 1993). This crop has become increasingly viable under UK conditions in recent years. It is eligible for Arable Area Payments, which removes some of the variability in its financial performance. Two types of variety, tall and late (late July) flowering, and short and early (mid-July) flowering, are available. Flowering typically lasts about 20–30 days. Experience at ADAS Boxworth, on a Hanslope clay soil, over four years is that later flowering sunflowers were more profitable than linseed. Sunflowers could become an important crop, up to 40 000 ha, in the south of the UK. This would be of benefit to bees by providing forage later in the summer.

It is unlikely that other major changes to arable cropping will result in particular value to bees. The area of linseed will probably decline, especially on heavy soils. The long-term prospects for lupins are still uncertain:

FIG. 2. The already large area of oilseed rape is likely to remain static.

FIG. 3. Sunflower could become an important crop in the south of the UK.

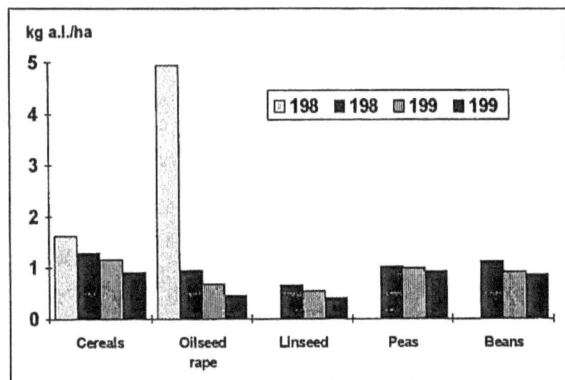

kg a.l./ha

Legend: 198 198 199 199

FIG. 4. Average rate of herbicide usage in Great Britain.

many agronomic issues need to be resolved before this is an attractive crop for farmers to consider. The areas of other crops such as evening primrose and borage are likely to remain small as a result of uncertain markets.

Recent changes in economics, attitude and the results of research, have led to greater scope to reduce the amount of herbicides and other pesticides used on arable land. In recent years the price of pesticides has increased while the value of crop output has reduced. This, coupled with an increased acceptance that a few weeds do not matter, has led to reductions in numbers of herbicide applications. Results from surveys of pesticide usage confirm that the amount of pesticide applied is decreasing[14]. In particular, herbicide use has declined since 1982 (fig. 4). This is to a large extent due to the increased activity of herbicides used, especially on oilseed rape, where TCA was widely used in 1982 at over 10 kg of active ingredient (a.i.) per hectare. This herbicide has been displaced by products which contain less than 0.5 kg a.i./ha. However, a significant amount of the reduction in pesticide usage is a reflection of farmers applying appropriate rates, at less than the full label rate, accepting that not all weeds need to be killed. This practice will continue and as a result a few more plants favourable for bee forage may survive in arable land.

Using set-aside

The introduction of set-aside into farm rotations enables us to consider what value this may be to bees and how the management of set-aside land may be tailored to specifically assist in improving bee forage.

Set-aside is not new to the UK, and management conditions have evolved over the years. The first European Commission (EC) set-aside scheme came into operation in the UK in July 1988[2]. The five-year scheme was voluntary, and in return for taking at least 20% of eligible arable land on a holding out of production, annual compensation payments of up to £222/ha of set-aside were available. Set-aside land could be rotated annually (rotational set-aside) or remain in the same place for up to five years (non-rotational set-aside). Land set aside had to be managed within defined conditions which included a requirement to establish a green plant cover, or, under certain circumstances, natural regeneration of the previous crop was permitted. The plant cover had to be maintained without the use of fertilizers or pesticides, except under certain exemptions. Cultivations were allowed to control weeds (provided a plant cover was re-established) or to establish a plant cover. Land set aside had to be cut[3].

In August 1991, another voluntary scheme, the EC one-year set-aside scheme, was introduced in the UK for one year only. Under this scheme, farmers were offered up to £121.20/ha plus a refund of co-responsibility levy paid on all cereals produced between 1 July 1991 and 30 June 1992. In return, farmers had to set aside at least 15% of their arable area used to produce specified crops for the 1991 harvest. Within this requirement, they also had to reduce their area of cereals by 15%. The management conditions of this scheme were similar to the five-year scheme with the exception that under the EC rules a green plant cover had to be sown (natural regeneration was allowed over the winter period if a green plant cover was sown in the spring)[4]. In 1991–1992, a total of about 167 000 ha (nearly 5% of the total cereal area) of land was set aside in the UK[5].

In May 1992 it was announced that set-aside would be part of the reform of the EC Common Agricultural Policy. Farmers entering the scheme were required to set aside 15% of their total area of cereals, oilseeds, protein crops and set-aside. Again, there had to be a plant cover (natural regeneration after cereal and herbage seed was allowed). However, after 1 May it was possible to cultivate the land to control weeds[6], and certain non-food crops could be grown on set-aside land. Around 550 000 ha has been set aside in the UK in 1993 under this scheme[8].

In 1993 the set-aside scheme was amended to allow for the introduction of non-rotational set-aside. In the UK, farmers who wish to opt for non-rotational set-aside are required to set aside 18% of the area of cereals, oilseeds (including linseed), protein crops and set-aside. The land must remain set aside for at least five years. Setting aside more than the minimum area, and mixtures of rotational and non-rotational set-aside, are allowed. Additionally, in 1993 the management rules of rotational set-aside were modified[7]. It is important to remember that exemptions from these management conditions may be sought for environmental reasons.

Since 1987, ADAS (an executive agency of the Ministry of Agriculture, Fisheries and Food (MAFF) and the Welsh Office) has been undertaking a series of experiments into set-aside management, funded by MAFF. The details and results have been fully reported[11, 12]. On set-aside the greatest number of plant species was found on the natural regeneration treatments, and the number of species was greatest on the lighter soil sites. In spring, following autumn establishment of rotational set-aside, there were 8.6 plant species/m^2 on the natural regeneration treatment on average across the five ADAS sites: by comparison there were 6.9 in Italian rye-grass and 4.6 in the continuous cereal plots. These numbers were generally reduced by the end of the summer.

Species found were mainly arable weeds such as brome (*Bromus* spp.), creeping thistle (*Cirsium arvense*), spear thistle (*Cirsium vulgare*), couch (*Elymus repens*), meadow grass (*Poa* spp.), groundsel (*Senecio vulgaris*), sow thistle (*Sonchus* spp.) and chickweed (*Stellaria media*). The species present were typical of the soil type and would be expected from the past cropping history of the sites. Some species listed as 'rare' by the Botanical Society of the British Isles were found in the experiments. Round-leaved fluellen

(*Kickxia spuria*) appeared at Boxworth and Drayton, and Venus's looking glass (*Legousia hybrida*) at Bridgets, despite the fields having a long history of arable cropping. In the set-aside fields on several farms, charlock (*Sinapis arvensis*) was the predominant cover in the first year, until it was cut. White clover has appeared naturally at two sites, both with a history of grass/clover leys in the rotation. No white clover has appeared in the majority of set-aside land which followed a long history of arable cropping. The value of natural regeneration set-aside to bee forage may therefore be somewhat limited in the majority of circumstances.

Although the number of species has increased over a four-year period on the non-rotational experiments, few of the species are of value to bees. There is also a conflict between the needs of farmers and bees, since some of the species are weeds to farmers but forage to bees. Many of the thistle and Compositae species are particularly aggressive weeds which farmers will not wish to see producing seed. Additionally, farmers see rotational set-aside as an important break crop in terms of weed numbers. It can provide an excellent opportunity to reduce the incidence of weeds such as brome and black-grass (*Alopecurus myosuroides*). Good control of these in set-aside can lead to reductions in the use of herbicides in following crops. This will mean much of the cover on rotational set-aside will be destroyed by herbicide, cutting or cultivation before it is of value as bee forage.

In 1992–1993, phacelia (*Phacelia tanacetifolia*) and mustard (*Brassica nigra*) were sown both in the autumn and spring in the rotational set-aside experiments. Satisfactory autumn establishment was achieved and mustard gave a good cover, but no flowers, before being killed off by frost early in the new year. By this date it had proved a useful cover in reducing potential nitrate leaching. Phacelia was slower to establish in the autumn and gave poor cover. However, it was more tolerant of frost. From the spring sowings, establishment was slow due to dry conditions and a very poor establishment of mustard resulted, especially on heavy soils. Although of value to bee forage, these cover crops are unlikely to be used on set-aside to any large extent. This reflects their cost implications and practicability. The greatest role for mustard is winter sown where a plant cover is required: this is unlikely to flower before it is killed by the frost. Neither cover is likely to be taken up on a large basis when sown in the spring. On cost grounds mustard is more attractive at about £10/ha, as phacelia seed is more expensive at about £50/ha.

Certain non-food crops can be grown on set-aside land. In 1992–1993, and for the foreseeable future, the most important of these are likely to be the annual crops of oilseed rape and linseed, and the perennial crops, for biomass production, of short rotation arable coppice: poplar (*Populus* spp.), willow (*Salix* spp.) and elephant grass (*Miscanthus* spp.). There is already a sizeable area of oilseed rape and the varieties grown are unlikely to increase the flowering period.

Set-aside land is therefore likely to be of restricted value to bee forage unless specifically managed for this purpose. However, there is great opportunity to use land not growing arable crops to enhance bee forage. Positive management of this land would be required, so what could be done?

Management of set-aside land for bee forage

First some objectives need to be set. Bee forage can be improved by provision of a greater area and distribution of beneficial plants, increasing the period of time that beneficial plants are available or supplementing forage area at times of the year when it is low. Greater areas of undisturbed land could be allowed, to offer nesting sites for wild bees. It is likely that non-rotational set-aside offers the greatest scope for benefit to bee forage, because there is less need to consider weed control aspects and greater scope to allow beneficial plants to develop.

Species that would be of most benefit to bees include white clover (*Trifolium repens*), red clover (*T. pratense*), crimson clover (*T. incarnatum*) and lucerne (*Medicago sativa*). Unfortunately these species are discouraged because they can be very competitive, thus reducing the number of species present, and also encourage nitrate leaching. The rules for rotational set-aside allow seed mixtures to contain up to 5% legumes, although derogations for higher contents

are allowed for organic farmers. On non-rotational set-aside clover and lucerne are specifically excluded, but other legumes could be included.

Two alternative solutions for non-rotational set-aside land exist. Firstly, sowing of plant species which are less aggressive, such as native cultivars of birdsfoot trefoil (*Lotus corniculatus*) and less vigorous species of vetches (*Vicia* spp.) should be considered. Alternatively, a derogation for inclusion of clovers could be sought from MAFF if there are clear environmental objectives, such as encouraging insect populations. Management to improve honey production would not be a legitimate objective as this would constitute agricultural production on the land, which is not allowed. Derogation requests are more likely to be viewed constructively if careful consideration is also given to the area required. For instance, only one hectare out of a 15-ha field could be appropriate. It is unlikely that MAFF would grant widespread derogations to the legume restrictions, but well-documented and sound, scientifically-based proposals to favour wild bees are likely to be viewed favourably, provided they are not linked in any way to honey production.

On rotational set-aside several options exist. The type of Tübingen mix and management proposed by Engels *et al.*[15] could offer one solution. This mix, containing 40% phacelia, 25% buckwheat (*Fagopyrum esculentum*), 7% white mustard (*Sinapis alba*), 6% coriander (*Coriandrum sativum*), 5% field marigold (*Calendula officinalis*), 5% black cumin (*Nigella sativa*), 3% red radish (*Raphanus sativus*), 3% cornflower (*Centaurea cyanus*), 3% common mallow (*Malva sylvestris*), 2% dill (*Anethum graveolens*) and 1% borage (*Borago officinalis*), provides a succession of forage but needs evaluation under UK conditions. Phacelia or mustard alone could be sown as covers on rotational set-aside land but would not provide forage for as long a period. On rotational set-aside it is essential that the implications of weeds, pests and diseases and the effects on yields of following crops are evaluated. To be grown on non-rotational set-aside land, these annual covers would require a derogation from MAFF.

Greater benefit to bees may come from non-rotational set-aside land. In the short-term beneficial species are unlikely to be found in high numbers. The first year is dominated by arable weeds, irrespective of whether a plant cover is sown or natural regeneration allowed, and the importance of these weed species is rapidly diminished by frequent cutting in the first year. Natural regeneration may eventually result in the right species composition, which will happen more quickly on light soils. It is more likely that low levels of wild flowers, appropriate to the location and soil type, would need to be sown to result in 'spiked' regeneration. Similarly a wild flower/grass mix could be sown. Wild flowers known to be favoured by bumble bees include bellflowers (*Campanula* spp.), buttercup (*Ranunculus* spp.), comfrey (*Symphytum* spp.), cranesbills (*Geranium* spp.), dandelion (*Taraxacum officinale*), foxglove (*Digitalis purpurea*), knapweeds (*Centaurea* spp.), white dead nettle (*Lamium album*), willow-herbs (*Epilobium* spp. and *Chamaenerion* spp.), and woundworts (*Stachys* spp.)[16].

The management of both situations can be adapted to suit many species of bees. After the more intense mowing regime in the first year, cutting in autumn or spring, rather than summer, is preferable. If possible large areas should not be cut at one time and a range of growth stages left[13]. The current UK set-aside management conditions require the plant cover to be cut between 15 July and 15 August, but it is possible to cut at a later time by obtaining a derogation from MAFF Regional Service Centres. Such applications need to explain the environmental objectives and state the cutting regime proposed.

Unfortunately most of these solutions incur some element of cost on behalf of the farmer. While sympathetic farmers will readily try to accommodate management practices suitable to a range of opportunities, many will need to see their extra costs covered. What these costs are will depend on the costs of the seed mixture and cultivation, the area required and, in the case of rotational set-aside, the effects on following crop yields. Under the rules of set-aside no profit, or benefit in kind, may be made by the farmer from set-aside land. Additionally production of honey is classed as agricultural production, which would not be allowed on set-aside land. Management to favour wild bees would be quite acceptable, provided there was no link with honey production. However, it may be necessary to persuade, or provide free, specific seed mixtures, rather than make payments to farmers.

General land management issues

Other activities that could favour bee forage will include the management of hedges, woodlands and uncropped areas. Set-aside around field margins could offer greater scope to manage the surrounding vegetation in a more sympathetic way and to carry out operations at almost any time of the year. Set-aside could be used to create new hedges which may result in smaller field sizes, provided these were still of a farmable size. Most farmers are already well aware of the need to consider bees when spraying fields and inform beekeepers accordingly. Field margins would help by keeping insecticides, other pesticides and fertilizer further away from the refuge and nest sites by creating a buffer between the field and beneficial areas.

The set-aside rules also allow for two metres near hedges or woods to be left uncut, and by derogation larger areas could also be allowed. This management would encourage scrub to develop, including brambles and herbage round the base, providing potential nest sites and possibly overwintering sites. In our set-aside experiments we have seen increased numbers of small mammals under cutting regimes which leave cover. Their disused nests, below-ground holes (fig. 5) or on the surface such as in grass tussocks, would be of benefit as nesting sites[1, 13]. Favoured overwintering sites for wild bees, such as north-facing banks on sides of woods, could also be encouraged. Often it is the part of the field on the northerly and eastern edges of a wood where the crop is poorest. By removing a 20-m strip as set-aside this could be managed to enhance the benefit to bees.

O'Toole[17] suggests that more rabbits could be the single most important contribution to wild bee populations. This is because the grazing pressure they exert can create and maintain areas of high floral diversity and bare ground around the burrow entrance where ground-nesting bees can nest. Rabbits will be favoured by undisturbed land and could increase as a result of set-aside. However, farmers and landowners

FIG. 5. Small mammal activity increases under less intensive cutting regimes, leaving holes suitable for ground-nesting bees.

will wish to ensure that any increase does not have a severe effect on cropped land.

Set-aside land could also be used to increase the area of woodland. At present the EC will not allow trees to be planted on set-aside land to receive Woodland Grant Scheme (WGS) establishment grants. It is the UK government's stated intent to negotiate for this to be changed, although farmers can already apply for an exemption to plant trees (without WGS or Farm Woodland Premium Scheme assistance) on set-aside land. Extra woodland needs to be sited to create a mosaic with existing habitats[13]. New and existing woodland could be managed specifically to the benefit of bees by allowing undisturbed vegetation with a mix of open rides, woodland edges, clearings and young plantations. This would provide nesting sites for wild bees and forage for bumble bees. New hedgerow plantings should include hawthorn (*Crataegus monogyna*), field maple (*Acer campestre*), and holly (*Ilex aquifolium*). Allowing bramble (*Rubus fruticosus*) and wild roses (*Rosa* spp.) to develop will also increase the value. Wild privet (*Ligustrum vulgare*) and buckthorn (*Rhamnus* spp.) are also valid choices[18]. For woodland and amenity plantings Norway maple (*Acer platanoides*), wild cherry (*Prunus avium*) and related species combined with sweet chestnut (*Castanea sativa*), horse chestnut (*Aesculus hippocastanum*) and lime (*Tilia* spp.) would offer large amounts of forage. Specific choice would depend on geographic location, soil and landscape type.

Further increases in the number of Environmentally Sensitive Areas (ESA) and Nitrate Sensitive Areas (NSA) which offer farmers payments to return land to grassland[9] will result in an increased area of land managed less intensively, which in turn will be of benefit to bees.

Further benefit to bees could result from the Habitat Scheme and Moorland Scheme. The UK Government has submitted proposals to the EC for these new schemes under the Agri-environment Regulation. The schemes will offer payments to farmers who undertake to withdraw land from production for 20 years for environmental reasons. Under the current rules, any such land would not count as set-aside under the Arable Area Payments Scheme. The UK Government is pressing the EC to reconsider this point in relation to the Habitat Scheme where land is withdrawn completely from production. The Habitat Scheme contains proposals for the following options: intertidal habitats (payments of £525/ha are available if the land is currently arable or temporary grass, £195/ha where land is currently permanent grass); water fringe habitats (£360/ha after arable or temporary grass, £240/ha after permanent grass); and existing set-aside habitats (£275/ha). Additionally there are two further options, for the creation of lowland heath (£380/ha) or damp lowland grassland (£335/ha) from arable land, if the EC rules enable land taken out of production under this scheme to count as set-aside under the Arable Area Payments Scheme. The Moorland Scheme is designed to encourage the conservation and enhancement of heather and other shrubby moorland vegetation. Proposed payments are £15 for each ewe eligible for Hill Livestock Compensatory Allowance by which their flock is reduced[9].

Suggested actions to improve bee forage

In writing this paper from the agriculturist's viewpoint, one striking feature is how little information is readily available about bees. We suggest that there are many involved in the agricultural industry who would wish to carry out management practices to favour bees. Other groups of interested parties have issued notes and information on how to manage land for specific objectives: examples include the Farming and Wildlife Advisory Group (FWAG), Royal Society for the Protection of Birds (RSPB) and the Game Conservancy Trust (GCT). All offer specific information on the management of land, including set-aside land, for the benefit of their particular interest, often birds in the cases listed. Similar information on bees would be well received.

There are good opportunities for promoting bee forage in the nineties, but they will not happen unaided. Information must be produced and disseminated to explain why bees are important, the types and management of habitats required, and which crops, cropping patterns and crop management practices are beneficial. The use of set-aside to increase the area of bee forage available gives new impetus to evolve management practices and land use of specific benefit to bees. In many cases this will also be complementary to the management required for other flora and fauna, allowing opportunity for joint publications. All land, whether set-aside or not, must be managed. Provision of bee forage, nesting and overwintering sites is another legitimate objective of land management. Promote the cause through farming journals and by lobbying interested groups to 'think bees'.

Acknowledgements

We are grateful to Ingrid Williams, IACR Rothamsted Experimental Station, and to David Alford, ADAS Cambridge, for their assistance in providing information on which to base this article. Financial support from the Ministry of Agriculture, Fisheries and Food (MAFF) is gratefully acknowledged.

References

The numbers given at the end of references denote entries in *Apicultural Abstracts*.

1. ALFORD, D V (1975) *Bumblebees*. Davis-Poynter; London, UK; 352 pp.   971/76
2. ANON (1988) *Set-aside, SA1*. Ministry of Agriculture, Fisheries and Food; London, UK.
3. ANON (1991) *Set-aside, SA1*. Ministry of Agriculture, Fisheries and Food; London, UK (revision 3).

4. ANON (1991) *One-year set-aside scheme, SAO3.* Ministry of Agriculture, Fisheries and Food; London, UK.

5. ANON (1991) *Set-aside: take up under one-year and five-year schemes.* Ministry of Agriculture, Fisheries and Food; London, UK; News Release 456/91.

6. ANON (1992) *Arable Area Payments: explanatory booklet AR2.* Ministry of Agriculture, Fisheries and Food; London, UK.

7. ANON (1993) *Arable Area Payments 1993/94: explanatory guide AR6.* Ministry of Agriculture, Fisheries and Food; London, UK.

8. ANON (1993) *Agricultural and horticultural census 1 June 1993: United Kingdom and England provisional results.* Ministry of Agriculture, Fisheries and Food Statistics; Guilford, UK; News Release Stats 164/93.

9. ANON (1993) *Agriculture and England's environment.* Ministry of Agriculture, Fisheries and Food; London, UK; News Release 266/93.

10. ANON (1993) *1993 Arable Area Payments applications.* Ministry of Agriculture, Fisheries and Food; London, UK; News Release 348/93.

11. CLARKE, J H (1992) *Set-aside. In* Clarke, J (ed) *BCPC monograph no. 50.* British Crop Protection Council Publications; Farnham, UK; 283 pp.

12. CLARKE, J H (1993) *Management of set-aside land: research progress.* Ministry of Agriculture, Fisheries and Food; London, UK; Leaflet AR13; 9 pp.

13. CORBET, S A; SAVILLE, N M; OSBORNE, J L (1994) Farmland as a habitat for bumble bees. *In* Matheson, A (ed) *Forage for bees in an agricultural landscape.* International Bee Research Association; Cardiff, UK; pp 35–46.

14. DAVIS, R P; THOMAS, M R; GARTHWAITE, D G; BOWEN, H M (1993) *Arable farm crops in Great Britain 1992.* Ministry of Agriculture, Fisheries and Food; London, UK; Pesticide Usage Survey Report 108; 89 pp.

15. ENGELS, W; SCHULTZ, U; RÄDLE, M (1994) Use of the Tübingen mix for bee pasture in Germany. *In* Matheson, A (ed) *Forage for bees in an agricultural landscape.* International Bee Research Association; Cardiff, UK; pp 57–65.

16. FUSSELL, M; CORBET, S A (1992) Flower usage by bumble bees: a basis for forage plant management. *Journal of Applied Ecology* 29: 451–465.

17. O'TOOLE, C (1994) Who cares for solitary bees? *In* Matheson, A (ed) *Forage for bees in an agricultural landscape.* International Bee Research Association; Cardiff, UK; pp 47–55.

18. ROBERTS, P (1994) What are the important nectar sources for honey bees? *In* Matheson, A (ed) *Forage for bees in an agricultural landscape.* International Bee Research Association; Cardiff, UK; pp 21–33.

www.ingramcontent.com/pod-product-compliance
Lightning Source LLC
Chambersburg PA
CBHW081410270326
41931CB00016B/3437